A GUIDE TO THE
PACKHORSE BRIDGES
OF ENGLAND

Goyts Bridge - Originally on an ancient salt route, this Derbyshire bridge was re-built higher upstream to avoid being drowned in Errwood Reservoir

A Guide to
the Packhorse Bridges
of England

by

Ernest Hinchliffe

CICERONE PRESS,
MILNTHORPE, CUMBRIA

© Ernest Hinchliffe 1994
ISBN 1 85284 143 5

A catalogue record for this book is available from the British Library.

ACKNOWLEDGEMENTS

I gratefully acknowledge the debt I owe to previous workers in this field, especially those local enthusiasts who have recorded their researches. Publications which I have found especially useful are detailed in the Bibliography.

Many packhorse bridges are isolated and evidence of the routes they served has often disappeared. In some cases I have speculated about these routes, largely from study of maps. Such guesses and the mistakes that go with them, are of course mine alone.

With a few exceptions I have visited all the bridges described, so compiling the guide involved a lot of travelling. Many librarians in Local Studies Libraries, responding generously to my written enquiries, helped me avoid wasted journeys. To them all I offer my sincere thanks.

To Alex Menarry, companion on many an expedition down muddy and overgrown lanes and through wet fields, following often obscure clues in search of "yet another bridge" - Thanks.

And finally, thanks to my wife Jean for her forbearance. Obsessions must be hard to live with.

Advice to Readers

Readers are advised that whilst every effort is taken by the author to ensure the accuracy of this guidebook, changes can occur which may affect the contents. It is advisable to check locally on transport, accommodation, shops etc but even rights-of-way can be altered and, more especially overseas, paths can be eradicated by landslip, forest fires or changes of ownership.

The publisher would welcome notes of any such changes

Front cover: Ings Bridge, just south of Pickering town centre

CONTENTS

PREFACE

In compiling this guide, a key question has been to decide "when is a packhorse bridge not a packhorse bridge?"

Of first importance is width. Before the eighteenth century, bridges were described either as "horse bridges" or "cart bridges" and the dimensions and manoeuverability of carts must have been important in distinguishing the one from the other. Having in mind the literature which describes the bewildering variety and size of carts, wains, trolleys, tumbrils and wagons; and making allowance for a wobbly vehicle drawn by an erratic animal on a rough surface, it is unlikely that a cart bridge would be less than six feet wide.

Age too is important. The packhorse transport system was slowly killed by the canals and by wheeled traffic on turnpike roads and since the first Turnpike Act was passed in 1663 and the last Turnpike Trust disappeared in 1895, it was a lingering death, spreading slowly from the south and east towards the north and west. It is unlikely that new packhorse bridges would be built in the closing stages of the turnpike period (packhorse trains were in any case using these new roads), so a building date before say 1800 seems appropriate.

But! Not all bridges built before 1800 and less than six feet wide, are packhorse bridges. There is an old three-arch bridge across the River Coln in the village of Bibury in Gloucestershire which has most of the attributes of a packhorse bridge, but is almost certainly a footbridge serving a row of seventeenth-century weaver's cottages. Also, some old and narrow bridges near farms were probably built to ease stock movement. Many bridges which are often described as packhorse bridges have either been widened to carry carts and later, motor traffic, or in a few cases have been dismantled and re-built in a new position.

These difficulties have persuaded me to classify the bridges into three groups. Group 1 are the most authentic, 6ft wide or less, built before 1800 and with known packhorse associations. Group 2 bridges are the ones which fall outside one or other of these limits, but are generally referred to as packhorse bridges. Group 3 bridges are more problematical. During my research, a reference such as "small", "narrow", "old", or "bridle bridge", would raise hopes of a genuine packhorse bridge. Some of them were. Others, though often very interesting were, in my opinion, not packhorse bridges. For those interested in deciding for themselves, these as it were "rejects", make up Group 3.

The guide is divided into three regions, North, Midlands and South, using the same boundaries as the Department of the Environment in their *Lists of Ancient Monuments*; and each Region is divided into counties. The Northern Region includes the Counties of Cumbria, Durham, Greater Manchester, Lancashire, Northumberland, North Yorkshire, South Yorkshire and West Yorkshire. The Midlands Region includes the Counties of Bedford, Cheshire, Derbyshire, Gloucester, Hereford and Worcester, Leicester, Lincoln, Norfolk, Northampton, Shropshire, Stafford, Suffolk, Warwickshire and West Midlands. The Southern Region includes the Counties of Avon, Cornwall, Devon, Dorset, Somerset and Wiltshire. The Regional and County boundaries are shown on Map 1 and the location of all the bridges described, on Map 2.

Where a bridge crosses a stream which is a county boundary, for example the River Dove which early in its course is the boundary between Derbyshire and Staffordshire, the alphabet is followed and for this case, the bridge described under Derbyshire.

Bridges are described in alphabetical order under the name of the nearest village. Where this is impossible because there is no village, I have used the name the bridge is generally known by. In all cases, in order to locate the bridge accurately, the name and group is followed by the number of the Ordnance Survey 1:50,000 Landranger Series map and the six-figure national grid reference thus:

Calva Hall (Group 1) 89: NY 059 265

Where the bridge is listed as an "ancient monument" by the DOE, this fact is noted and the county number given.

All the measurements quoted are mine and they do not pretend to be highly accurate. I used a tape measure for width and height of parapet. For the span, I strode (one stride for one yard) between the arch springings. On multi-span bridges, I measured between the outer springings of the outer arches.

I feel sure that there are packhorse bridges in England which have escaped my researches. I would be glad to know of them. If the guide stimulates people to search out their local bridge, or leave the beaten track to find one whilst on holiday, it will have served its purpose.

PACKHORSE TRANSPORT

Mankind's transition from hunter-gatherer to settled farmer, prompted a new motive for travel. Instead of the never-ending quest for food, he travelled to trade the artefacts produced by his increasing skills. Langdale axe-heads are good evidence. These lovely stone-age implements, mined and roughly shaped in the heart of the Lake District and polished in coastal settlements, have been found the length and breadth of England. Since stone axes are heavy and horses were domesticated in prehistoric times, it seems reasonable to speculate that ancient man used horses as beasts of burden. Early written evidence in support is provided by a Latin author, Deodorus Siculus, writing in the first century BC before the Roman occupation of Britain. He describes the carriage of Cornish tin by packhorse from a landing place on the Atlantic coast of Gaul, to Narbonne on the Mediterranean.

The Romans, whose military success depended to a large extent on the superb network of military roads which they built to cope with wheeled vehicles, had limited need for pack animals. In the chaos following their departure, roads decayed and trade declined.

Not until law and order began to return under such "Dark-Age" Kingdoms as Northumbria, Mercia and Wessex, did trade expand again and it is probable that the first long distance packhorse transport system was organised then to carry that vital preservative, salt. From evaporating pans on the south and east coasts (Domesday Book identifies over one thousand of them) and from Droitwich and the Cheshire "wiches", salt was carried throughout the country. Map 3 illustrates some of the salt routes and the main coastal areas where salt pans have been found.

After the Norman conquest, the prosperity of the early Middle Ages was based on wool, which by 1300 provided nearly all England's export earnings. The trade peaked in 1305 when 46,382 sacks, each of 364lbs were exported. *Note* that historical accounts of the Medieval wool trade seem to agree on 364lbs as the weight of a woolsack, whereas 3cwt (336lbs) seems to be about the maximum carried by one packhorse. I speculate that the sacks were either split

8

for transport, or the wool was made up into woolsacks at the port.

Foremost in this lucrative trade were the monasteries of the Cistercian Order, wool growers *par excellence*. The widespread distribution of their main establishments is shown on Map 4. No doubt water-borne and wheeled transport was used where possible, but much wool was carried by packhorse.

The dissolution of the Monasteries in the mid sixteenth century had an important effect on transport. However reluctantly, religious establishments had accepted some responsibility for the repair of roads and bridges and no organisation took their place. It seems probable that in late Tudor times, the roads were in a worse state than they had been for two or three centuries before.

After the chaos of the Civil War, the period from the middle of the seventeenth century to the end of the eighteenth saw the peak of packhorse transport, especially in the north of England. Trains of up to forty animals, the leading horse adorned with bells, travelling perhaps fifteen miles a day, each carrying between 2 and 3cwt, formed the core of the long distance trade. To get a feel for the increase in commerce between those days and the present, consider a 36-ton juggernaut delivering a load from Carlisle to London in a day. More than 300 horses would be needed to carry the same load and would take three weeks to deliver it.

The pack trains can never have had an easy passage. Apart from difficult river crossings, there were outlaws and highwaymen, Civil War, Enclosure Acts and the sometimes atrocious weather (there are stories of whole packhorse trains being frozen to death in Britain's "little ice age", at its worst in the 1690s). They also had to contend with the roads. It is difficult to appreciate just how awful were many English roads for much of the long period between the departure of the Romans and the building of the turnpikes. Perhaps only farmers and ramblers now appreciate the clinging properties of wet clay, or the porridge that peat can become under too many feet.

At the peak of the trade, the variety of freight carried by packhorses was enormous. Wool and cloth to the ports and the great inland markets such as Stourbridge; corn, carried for local distribution by licensed "Badgers"; lead and iron ores to the smelt mills; coal and charcoal; chert from Derbyshire quarries to the Potteries; Sheffield cutlery to London; lime, both for building and to sweeten acid pastures; even goose feathers: the list is almost endless.

Not all these cargoes were long-distance. Metal extraction, such as iron in South Yorkshire, tin in the south-west, or lead in Derbyshire and the Yorkshire Dales, generated much local traffic. In *Old Yorkshire Dales*, A Raistrick, writing of the smelt mill at Yarnbury above Grassington, and considering ore, coal and general stores into the mill and lead from it to market, estimates that this one local industry generated 16,000 pack loads per annum.

Perhaps the most surprising cargo is described by Defoe in his *A Tour through the whole Island of Great Britain*, published between 1724 and 1727.

> "This River Derwent is noted for very good salmon, and for a very great quantity, and trout. Hence, that is, from Workington at the mouth of this river, and from Carlisle, notwithstanding the great distance, they at this time carry salmon (fresh as they take it) quite to London. This is perform'd with horses, which changing often, go night and day without intermission, and, as they say, very much out-go the post; so that the fish come very sweet and good to London, ..."

I wonder! Perhaps the salmon were not so "very sweet and good" as we would nowadays expect.

Goods were normally carried in panniers slung either side of the horse from wooden pack-frames and to allow clearance for these, the parapets were very low or non-existent when packhorse bridges were first built.

Various breeds of horse were used. One favourite was a sturdy animal derived from a German hunter called a Jaeger, another was the Galloway from south-west Scotland. In the hill districts of Dartmoor, the Yorkshire Dales and the Lake District, locally bred fell ponies were used.

Map 5 shows the widespread distribution of "Packhorse" and "Woolpack" inns and public houses and worth pointing out is the long line of Woolpack Inns leading from the north-west towards London and onwards to the channel ports. Also the dense concentration of both Woolpack and Packhorse establishments in the industrial area of the Southern Pennines. Other relics of packhorse days are confined mostly to the hills. Causeways and Holloways, stone guide posts, packhorse bridges, place-names and bridleways or green lanes, together can often be used to trace old packhorse routes over surprisingly long distances.

Between the departure of the Romans and the building of turnpike roads and canals more than a thousand years later, pack animals provided the only really viable long-distance overland transport. It is a humbling thought that the combined age of turnpikes, canals, railways, motorways and aircraft is not much more than 300 years. The packhorse bridges which remain to delight us, built on a reassuringly human scale and often tucked away in unspoiled rural corners, serve as reminders of the long period when transport was a more leisurely affair.

GLOSSARY OF WORDS WITH PACKHORSE ASSOCIATIONS

BADGER
A Badger was an individual pedlar or trader licensed ("badged") in the seventeenth century to carry corn from an important market to sell at smaller markets or to isolated communities. There are a few old tracks which still have "Badger" as a name element. A "Badger Way" crosses Barningham Moor between Bowes and Marske in North Yorkshire with the added interest of a guide post at Grid Ref: NZ 064 077 still legibly inscribed "Badger Way Stoop". Parts of another old road between Beamsley and Harrogate are called "Badger Gate". There are other examples.

BANNISTER
A bannister was a pannier (q.v.) with a trap-door base used for carrying heavy and amorphous loads such as lime, manure, coal or metal ore.

BRIDLE
Said of a bridle-path, bridle road, bridle-way or bridle-bridge. Suitable for horses but not for wheeled vehicles. Longshaw packhorse bridge (q.v.) in Derbyshire is often referred to as a bridle-bridge.

BROGGER
A brogger (from broker?) was a middleman acting between peasant farmers and merchants. In Yorkshire he specialised in the trade and carriage of wool; fleece to the spinner, yarn to the weaver, cloth to the fuller etc. An act of 1552 abolished "Broggers, Ingrocers, Woolgatherers and sondrie other persons", probably because they were threatening the vested interests of more powerful merchants. Because in the Halifax district, merchant and brogger were often the same man, the Halifax Act of 1555 exempted the district from the abolition. I am still searching for a "Brogger" place-name.

CARROWAY
Derived from Carrier-Galloway (q.v.) and used to describe packhorses in the Weardale area, especially those used in the lead mining industry.

A flagged causeway near Penistone, South Yorks

CAUSEWAY

An Old French word *caucie* meaning an embankment or raised track has been corrupted to "causeway" or "causey", used to describe a path built up to save a traveller from foundering in the mire. At a time when Parishes were (reluctantly) responsible for repairing roads, they often dodged the issue by building a causeway over the stickiest patches. As J.Crofts writes in *Packhorse Wagon and Post* - "Thus they made a causeway because they had left the road soft and continued to leave the road soft because they had made a causeway". By an Act of Parliament of 1691, roads to market towns were to be 8ft wide and "Horse Causeys" not less than 3ft wide. Many fragments of causeway remain on old packhorse ways, often paved with stone "flags", especially in the South Pennines. A good length can be found near Penistone (South Yorkshire) at Grid Ref: SE 248 048.

CROOK

A U-shaped, lightweight wooden framework, fixed to the packsaddle and used for carrying light but bulky produce such as hay, especially in the south-west. There is a crook in the Torquay museum.

13

*A holloway on "Jagger Lane" near Gilling West
north of Richmond, North Yorks*

CROSS

In the Middle Ages, wayside crosses (as opposed to churchyard or market crosses) were often erected to guide the traveller. The best are on the high ground of the Pennines and North York Moors, though there are many more surviving base sockets than crosses. A good example is "Edale Cross" (Grid Ref: SK 080 861) at a high point on an old road between Cheshire and Yorkshire, some parts of which are still known as the "Monk's Road".

GALLOWAY

A popular packhorse breed from south-west Scotland. "Galloway Gate", a long distance packhorse route from Lancaster to the north may commemorate the breed. Also one opinion suggests that Gallox Bridge at Dunster, Somerset (q.v.) derives from Galloway.

GATE

Comes from the Old Norse word *Gata* meaning a street or road. Common enough as part street-name in some Medieval towns, it is also used on some ancient cross-country routes - eg. Morpeth Gate leading to West Burton packhorse bridge, North Yorkshire (q.v.) and Cut Gate, from Penistone to Slipperystones packhorse bridge, Derbyshire (q.v.).

GRANGE
The outlying farm(s) of monastic establishments, especially the Cistercians. Many modern "Grange Farms" are founded on the site of an old Monastic Grange and are very numerous. For example there are fifteen "Grange Farms" within five miles of the ruins of Louth Park Abbey (Cistercian) in north Lincolnshire. Any remaining fragments of unmodernised track between Abbey and Grange would undoubtedly have seen many packhorses.

HOLLOWAY
A pathway on soft ground under centuries of pounding from horses hooves and erosion by water, especially on hillsides, tended to wear down into a deep trench or "holloway", sometimes "hollowgate", modified to "holgate". Many remain, both on the ground and in place-name. The road through that district of north London which houses the women's prison had to have a stone foundation as long ago as 1417, and Gilbert White, in *The Natural History of Selborne* describes the road between Selborne and Alton as being a holloway 16 or 18ft deep. Two (of many) examples with packhorse associations are in Derbyshire at Grid Ref: SK 251 801 and in North Yorkshire at Grid Ref: NZ 046 977.

Jagger's Lane - Hathersage, Derbyshire

JAGGER

The packmaster came to be called a "Jagger" and two origins are possible. The word may derive from a northern dialect word "Jag", meaning a load, or, more romantically it is a corruption of *Jaegar*, a German hunter and one of the favourite breeds of packhorse. There is still a "Jagger Lane" in Hathersage, Derbyshire, and another a few miles north of Richmond in North Yorkshire. Also in Derbyshire at Grid Ref: SK 153 872, an important packhorse road between Cheshire and Yorkshire winds its way around "Jaggers Clough". There are other examples.

NICK

In hilly areas, a notch in a ridge or edge which breaks the skyline when seen from the valley is sometimes called a "nick" and was used by the packmaster to help guide his train through the hills. "Scarth Nick" (Grid Ref: NZ 473 003) where the Hambledon Drove climbs steeply on to the North York Moors; "Windgate Nick" (Grid Ref: SE 070 471) indicating the packhorse road on to Rombalds Moor from Addingham; and "Oldgate Nick" (Grid Ref: SJ 995 762) on a salt road from Cheshire to Buxton; are three examples.

PANNIER

The name given to the basketwork or leather container slung either side of the pack saddle.

SALT

The carriage of that essential commodity salt, has left many a "Saltersford", "Saltergate", "Salter's Lane", "Saltersbrook" and similarly named places. The best concentration known to me is on the old salt road over the Bowland Fells in Lancashire. On the descent from "Salter Fell" (Grid Ref: SD 645 592) towards the River Lune at Hornby (Grid Ref: SD 585 684), "High Salter Close", "High Salter", Middle Salter" and "Lower Salter" follow in quick succession.

STOOP

Under legislation in the reign of William and Mary (1697), Justices of the Peace were ordered to erect guide posts at important crossroads. Called "stoops", many of them remain in situ in the pennines from the Peak District northwards. Usually a square post five or six feet high, the direction was carved on the side facing the destination. Not all districts obeyed the 1697 order, which had to be repeated in 1733. A good selection with Grid References is given in *Packmen, Carriers and Packhorse Roads* by D.Hey.

STREET
From the Anglo-Saxon *straet* and very often applied to a Roman Road - eg. Watling Street.

SUMPTER
A very old word which can mean a packhorse, or a packmaster, or even his saddle bag. "Sumpter Yard", just east of St Albans Cathedral, is the place where supplies were once unloaded for the Abbey.

WAY
From the Old English *weg*. Modern usage would probably be "road", but the old word "way" persists in, for example, highway, wayside, right-of-way etc. "Abbot's Way" across Dartmoor between Buckfast and Tavistock Abbeys is an example of this usage.

Sumpter Yard, near St Albans Cathedral,
where packhorses were unloaded for the abbey

Map No. 1

REGIONAL, COUNTY AND NATIONAL PARK BOUNDARIES

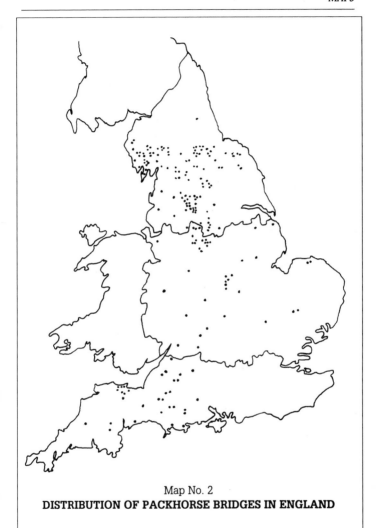

Map No. 2
DISTRIBUTION OF PACKHORSE BRIDGES IN ENGLAND

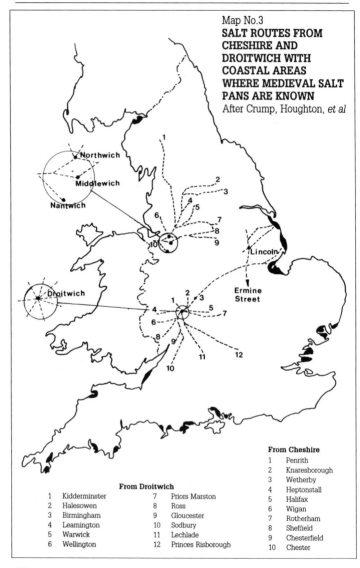

Map No.3
**SALT ROUTES FROM
CHESHIRE AND
DROITWICH WITH
COASTAL AREAS
WHERE MEDIEVAL SALT
PANS ARE KNOWN**
After Crump, Houghton, *et al*

From Droitwich

1	Kidderminster	7	Priors Marston
2	Halesowen	8	Ross
3	Birmingham	9	Gloucester
4	Leamington	10	Sodbury
5	Warwick	11	Lechlade
6	Wellington	12	Princes Risborough

From Cheshire

1	Penrith
2	Knaresborough
3	Wetherby
4	Heptonstall
5	Halifax
6	Wigan
7	Rotherham
8	Sheffield
9	Chesterfield
10	Chester

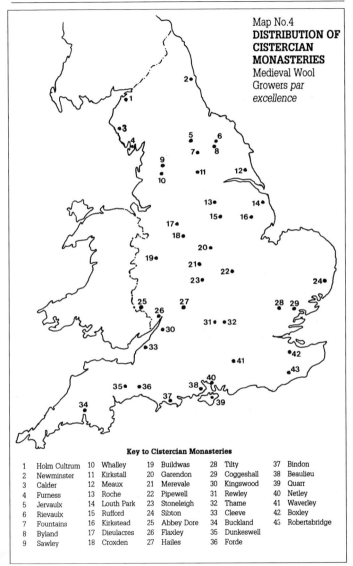

Map No.4
**DISTRIBUTION OF
CISTERCIAN
MONASTERIES**
Medieval Wool
Growers *par
excellence*

Key to Cistercian Monasteries

1	Holm Cultrum	10	Whalley	19	Buildwas	28	Tilty	37	Bindon
2	Newminster	11	Kirkstall	20	Garendon	29	Coggeshall	38	Beaulieu
3	Calder	12	Meaux	21	Merevale	30	Kingswood	39	Quarr
4	Furness	13	Roche	22	Pipewell	31	Rewley	40	Netley
5	Jervaulx	14	Louth Park	23	Stoneleigh	32	Thame	41	Waverley
6	Rievaulx	15	Rufford	24	Sibton	33	Cleeve	42	Boxley
7	Fountains	16	Kirkstead	25	Abbey Dore	34	Buckland	43	Robertsbridge
8	Byland	17	Dieulacres	26	Flaxley	35	Dunkeswell		
9	Sawley	18	Croxden	27	Hailes	36	Forde		

Map No. 5
**DISTRIBUTION OF 'WOOLPACK' AND 'PACKHORSE' INNS
AND PUBLIC HOUSES**

• 'Woolpack'

X 'Packhorse'

A NOTE ON THE HISTORY AND CONSTRUCTION OF BRIDGES

From the earliest times, small streams were no doubt bridged by the simple expedient of throwing a tree trunk or stone slab across. A good example of this type of "structure" (known as a clam bridge from an Anglo-Saxon word meaning twig or stick) remains at Wycoller in Lancashire. For wider spans, a series of slabs on intermediate supports served. The best-known of these multi-span clapper bridges are on Dartmoor and they were undoubtedly used by packhorses there. Though sometimes described as prehistoric, it is unlikely that any clam or clapper bridges which exist now were in place before the Middle Ages.

For larger rivers, a spot had to be found where depth and flow were low enough and the river bed firm enough to permit wading across - a ford. From Bideford in Devon, through Hereford, Oxford and Salford to Gainford in County Durham, many scores of place-names recall this early method of crossing and it may be that our ancestors chose these places with such skill that they were in continuous use until the settlements themselves acquired names.

The Romans, although they are generally credited with having introduced the semi-circular arch to Britain, have left no arched bridges. The assumption is that they placed wooden spans on stone abutments and piers and these, like many a later wooden bridge, have rotted away.

The replacement of fords, ferries and simple slabs by arched-stone bridges began before the Norman conquest. There are references to "a bridge of stone arches" being built at Winchester in the ninth century.

In the early Middle Ages, the great abbeys and monasteries built bridges as works of piety to relieve suffering travellers, though the wool trade, from which they greatly profited may have prompted them too. Town corporations and Guilds were other builders. In at least one instance royal survival provided the motivation. It is said that Queen Matilda (or Maud), who nearly drowned whilst negotiating the ford across the River Lea at Stratford in east London, had Bow Bridge built there.

In general, the responsibility for building and repairing bridges was one that nobody wanted and disputes were evident as early as 1215. Paragraph 21 of the Magna Carta is translated as: "No township or subject shall be compelled to make bridges at river banks, except those who by ancient usage are legally bound to do so."

When repairs were needed and "ancient usage" could not be determined, an inquisition was held to decide who was responsible. The verdict often seems to have been - No one! The King would then issue a "Grant of Pontage", which in effect was a licence to collect tolls in return for meeting the cost of maintenance.

In raising funds, the church was more enterprising than the State. On many bridges, chapels were built where travellers could offer alms as well as prayers to aid a safe journey and a few "Bridge Chapels" remain. They also sold indulgences. By the end of the Middle Ages, the going rate was 40 days remission of purgatory in return for money, material or labour given for bridge repair.

To mark the importance of bridges in Medieval life, the French even had a monastic order of Bridge Friars devoted to building and repairing them. Founded by St Benezet, the Friars built the famous Pont d'Avignon. The Order came to England, but their activities were limited and seem to have been confined to the Diocese of Exeter where, in the twelfth century they were granted a charter of protection and issued with Letters of Indulgence to help them raise money.

Medieval bridges were built using a variety of arch-shapes. The Roman or semi-circular arch was in use from the earliest times. Though very strong, it has the disadvantage that the longer the span, the higher the hump must be, until the steep climb itself becomes a hindrance to progress.

A segmental arch-shape reduces the height for a given span but requires stronger abutments to contain the greater splaying forces.

The problems caused by a high hump or the need for strong abutments are reduced by building several arches springing from intermediate piers, though these have the disadvantage that the piers must withstand the full force of the stream in flood, a need which leads to the cutwater. Generally triangular, less often rounded, the cutwater smooths and diverts the force of the water like the prow of a boat and often serves a secondary purpose by being continued upwards to provide a pedestrian refuge. The packhorse bridge in the village of Anstey near Leicester has five semi-circular arches. Each pier has both upstream and downstream cutwaters and above each cutwater is a refuge. Eight refuges in a total span of only 57ft seems generous. Perhaps the packhorse trains passing through Anstey were particularly inconsiderate to pedestrians.

The pointed arch, which can be seen on the Medieval packhorse bridges at Charwelton in Northamptonshire and Sutton in Bedfordshire amongst others, came into use in the fourteenth century and is rare after the fifteenth century. It is thus a good indicator of building date. Another Medieval technique was the use of ribs, which have the merit of using less high quality dressed stone and thus saving money. Short slabs, acting as beams were then laid across the ribs. A good example is Hunter's Sty bridge across the River Esk at Westerdale in the North York Moors.

The Tudors have given their name to an arch-shape: the Tudor or four-centred arch, which matches the period of church architecture known as perpendicular. The shape, though pretty, and common in churches, is rare in bridges. The old bridge at Cromford in Derbyshire is one, although the shape can be seen only on the downstream side. The bridge has been widened on the upstream side using a semi-circular arch-shape.

Several packhorse bridges have an arch-shape which approximates to a semi-ellipse, the only other shape in (not very common) use. The various shapes of arch, all having the same span, are shown in Fig. I (p.28) and the generally accepted names for the parts of a bridge are shown in Fig. II (p.29).

Irrespective of arch-shape, packhorse bridges are of two kinds. There are the small, single arch bridges deep in the hills, of what has been called "rustic" construction, which looks as though a dry stone waller had built them whilst seeking a change from his normal work. For wider spans, with the increased forces involved, more care, expertise and dressed stone were needed.

The Tudors are not only known for a particular arch-shape of course. They forced through many major changes in the traditional Medieval way of life. In 1530, counties were made responsible for the upkeep of main bridges and parishes for small bridges, probably in recognition that Grants of Pontage (the recipient seemed less ready to undertake repairs than to collect the tolls) and the sale of indulgences, were no longer effective. By 1540 the dissolution of the Monasteries was complete and any lingering sense of Church obligation for the upkeep of bridges disappeared. A little later, in 1548, Edward VI, by abolishing the guilds, removed another group with traditional responsibilities for bridge maintenance.

Many packhorse bridges can be firmly dated to the Stuart period, either from records, or because the bridges themselves are dated. Indeed it seems probable that except for the known Medieval examples, most of the packhorse bridges which remain were built between about 1660 and 1760, that is, between the middle of the Stuart and the middle of the Georgian periods.

Throughout history, bridges have been washed away by floods or destroyed by ice (it still happens), though no recent flood has been as disastrous as that in 1771 when all the bridges across the River Tyne except Corbridge were washed away, or the Norfolk floods of 1912 which destroyed eighty bridges.

Not all hazards were or are, natural. In times of strife, a bridge can be a prize like those at Arnhem and Nijmegen during the last war. During the English Civil War, with the Royalist forces besieged at Newark, two nearby bridges across the River Trent at Kelham and Muskham were destroyed and replaced by bridges of boats. There are many other examples of damage or destruction by both Royalist and Parliamentary forces.

Any kind of civil disturbance puts public property at risk. During the unrest in the reign of George IV, damage to a bridge was a transportable offence and several bridges in Dorset are still decorated by public notices which read:

DORSET

Any Person wilfully INJURING

Any part of this COUNTY BRIDGE

Will be Guilty of a FELONY and

Upon conviction liable to be

TRANSPORTED FOR LIFE

By the Court

7&8 GEO 4 C50 S13   T FOOKS

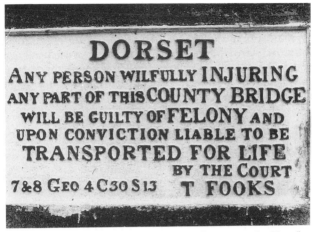

DORSET

ANY PERSON WILFULLY INJURING
ANY PART OF THIS COUNTY BRIDGE
WILL BE GUILTY OF FELONY AND
UPON CONVICTION LIABLE TO BE
TRANSPORTED FOR LIFE
BY THE COURT
7&8 GEO 4 C30 S13   T FOOKS

About 16 Dorsetshire bridges are fitted with "Transportation Plates".
This one is on Benville Bridge about a mile upstream from
Rampisham packhorse bridge

Many old bridges, large as well as small, are known as "Devil's Bridge". The original legend seems to come from a "Devil's Bridge" over the falls of Reuss in Switzerland. The Devil destroyed all other bridges over the river, but agreed to leave this one if the local Abbot would give him the soul of the first living thing to cross. The Abbot threw bread across, which was chased by a dog, so the Devil, wanting a human soul, was cheated. The variation in this country, which attaches to Devil's Bridge near Aberystwyth and Devil's Bridge at Kirkby Lonsdale, has the bridge built by the Devil to help an old woman retrieve her cow which, having strayed, was trapped on the wrong side of the river by a flood. The bargain between the Devil and the old woman is the same as at Reuss, and again the Devil is cheated by the same trick with bread and dog.

As with the Devil, many bridges in the country are wrongly described as "Roman". My own feeling is that after a couple of generations, the building of the bridge was forgotten. It then needed only a few early antiquaries, who liked to describe any old masonry as "Roman", to record their speculation, for the bridge to become "Roman".

How is it that so many bridges have survived hundred of years of neglect, floods and downright vandalism? First an arch is a very stable construction, and second, they were and are, necessary for any kind of civilised life and so neglect and damage eventually gets repaired. It is to be hoped that now packhorse bridges are unnecessary, their beauty and historical interest will ensure their survival for several more hundred years.

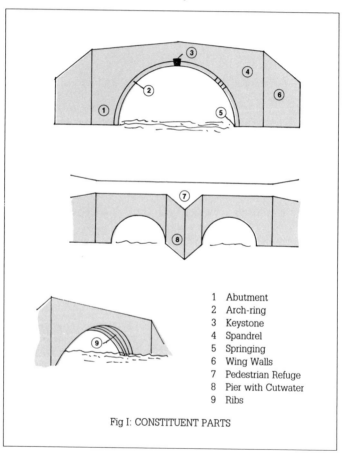

1   Abutment
2   Arch-ring
3   Keystone
4   Spandrel
5   Springing
6   Wing Walls
7   Pedestrian Refuge
8   Pier with Cutwater
9   Ribs

Fig I: CONSTITUENT PARTS

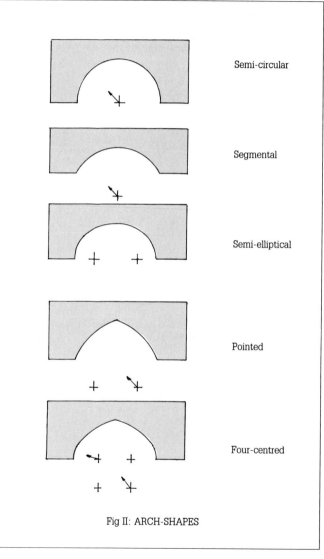

Semi-circular

Segmental

Semi-elliptical

Pointed

Four-centred

Fig II: ARCH-SHAPES

THE BRIDGES OF THE NORTHERN REGION

CUMBRIA
Ashness Bridge
Barbon
Birks Bridge
Bleabeck Bridge
Boot
Burnbanks
Calva Hall
Crosby Ravensworth
Dent Head
Drigg
Eskdale
Far Easedale
Furness Abbey
Gaisgill
Hartsop
Hawk Bridge (near)
High Sweden
Horrace
Lind End Bridge
Monks Bridge
Orton
Rosgill
Sadgill
Smaithwaite
Stainton
Stang End
Stockley Bridge
Throstlegarth
Ullock
Walna Scar
Wasdale Head

Watendlath
Water Yeat Bridge
Wet Sleddale
Winster

DURHAM
Deepdale
Egglestone Abbey
Headlam
Ketton Hall
Thwaite
West Hope

GREATER MANCHESTER
Prestolee

LANCASHIRE
Bleasdale
Catlow Bottoms
Croston Town Bridge
Higherford Old Bridge
Hodder - Old Lower Bridge
Wycoller

NORTHUMBERLAND
Ovingham

NORTH YORKSHIRE
Aldbrough
Askrigg
Barton
Baysdale Abbey

Birstwith
Boltby
Caldbergh
Clapham
Croft
Danby
Dob Park
Dousgill Farm
Glaisdale
Hampsthwaite
Harrogate
Hubberholme
Ivelet
Knox
Ling Gill
Linton in Wharfedale
Marske - Pillimire Bridge
Masham - Swinton Park
Newsham Lodge
Pickering
Ravenseat
Ravensworth
Romanby
Sinnington
Sowerby
Stainforth
Stokesley
Telfit
Thornsgill
Thornthwaite
Uckerby
Wath
West Burton
Westerdale
Yockenthwaite

SOUTH YORKSHIRE
Longshaw
Oxspring
Rivelin Valley
Unsliven
Wharncliffe Side
Whiston

WEST YORKSHIRE
Alcomden
Beaumont Clough
Bingley - Beckfoot Bridge
Clayton West
Crimsworth Dean
Eastwood
Hawks Clough
Haworth - Long Bridge
Hebden Bridge
Hebden Bridge (Foster Lane)
Hudson's Bridge
Lower Strines
Lumb Foot
Marsden (Closegate)
Marsden (Village)
Oxenhope
Oxygrains
Ragby Bridge
Ripponden
Standedge
Wakefield

INTRODUCTION TO THE NORTHERN REGION

Of the twelve counties which make up this Northern Region, eight are fortunate in having surviving packhorse bridges. Indeed three of the eight; Cumbria, North Yorkshire and West Yorkshire, possess more than half of all the bridges described in the guide. This is not accidental; until the coming of the turnpikes and railways, the remote mountain fastnesses of Cumbria and the Pennines were difficult to penetrate by anything on wheels.

The region is blessed with four National Parks, each with its own distinct character, two of them with many interesting packhorse bridges.

Nowhere more than in the Lake District are there packhorse bridges in such majestic settings. Throstlegarth and Stockley Bridge must be traversed by thousands of ramblers and fell-walkers each year on their way to the high hills. W.G.Collingwood, writing in 1928, made a convincing case that all the Lake District packhorse bridges were built between 1660 and 1760 and apart from the handful of known Medieval bridges further south, the case is probably true for the rest of the country.

Kendal, on the eastern fringe of Lakeland proper, was a very busy junction for packhorse transport. Regular weekly pack trains left the town for London, Glasgow, Wigan, Settle, York, Appleby and Barnard Castle, as well as most of the significant towns within the district. A major cargo from the town was woollen stockings. Nearly 30,000 dozen, mostly knitted in the surrounding countryside, were being transported from Kendal every year by the end of the eighteenth century.

The monasteries had a great impact on the Lake District in Medieval times. In what is now Cumbria, a dozen abbeys had been established by the end of the twelfth century, of which three were Cistercian, including the largest and richest, Furness Abbey. The monks were important wool growers and traders and early exploiters of the rich haematite iron ores of west Cumberland and Furness. They also traded salt, obtained from evaporating pans on the coast.

Later, in the time of Queen Elizabeth the first, the Society for the Mines Royal was established in Keswick to exploit the copper, silver

Ashness Bridge near Watendlath, Cumbria
Bleasedale Bridge, Bowland, Lancashire *(A.Menarry)*

and lead in the hills to the west of Derwentwater and in the Newlands valley. The ore was carried by packhorse to "Copperheap Bay" (at Grid Ref: NY 254 217 and still so named) on the shore of the lake and shipped to a smelter at Brigham, just north-east of Keswick.

Celia Fiennes, travelling in 1698, left a clear picture of Lakeland roads. Riding from Kendal to Bowness-on-Windermere, she complains that ..."here can be noe carriages, but very narrow ones like little wheelbarrows..." She also gave a good description of packhorses and their loads: "they also use horses on which they have a sort of pannyers some close some open that they strew full of hay turff and lime and dung and everything they would use, and the reason is plaine from the narrowness of the lanes."

Many Lakeland passes still show traces of the well-engineered zig-zags used to ease the gradient for the horses. The Kentmere side of the Nan Bield pass and the Langdale side of Stake Pass are good examples. The road over those adrenalin-generating passes for motorists, Wrynose and Hard Knott, was the last route in England to be regularly travelled by packhorses.

The Yorkshire Dales National Park also has its own very individual character, typified perhaps by a landscape of drystone walls and field barns. Here too the Cistercians were dominant in the countryside before the Dissolution. A million acres of the superb grassland of the Craven limestone country were owned by Fountains Abbey and used mainly for raising sheep. Later, lead was a major cargo, either as ore from the mining areas in Wharfedale and Swaledale to local smelt mills, or as the finished product to the lead markets at Richmond and Boroughbridge.

The North York Moors National Park is smaller and quite different from either the Lake District or the Yorkshire Dales. The bridges are fewer, but the three described in this guide, which cross the River Esk, are outstanding examples. As in the Yorkshire Dales, the high ground is crossed by old highways. The "Hambledon Drove" which follows the western scarp of the Hambledon Hills on its way from York to the crossing of the River Tees at Yarm (the Tees has been bridged at Yarm since the early thirteenth century); the "Monk's Causeway", joining Whitby Abbey and Guisborough Priory; "Saltergate" along which salt was carried from Pickering to

---

Ivelet Bridge, Swaledale, Yorkshire

Whitby for preserving fish; are three of the most important.

Locally, the packhorse was known as a Pannier Pony and the paved causeways which they used, "Pannerways". Stretches of pannerway can be found alongside the road up Glaisdale and in the vicinity of Commondale. The packhorse cargo peculiar to the North York Moors was Alum, used in the making of vegetable dyes, the mining of which has scarred the coast both north and south of Whitby. It was a Royal monopoly of the Stuarts in the mid-seventeenth century.

To the north-east of the main spine of the Pennines, lies the Northumberland and Durham coalfield and the industrial areas of Tyne and Wear and Tees-side which the coal helped to create. Carried by packhorse and wagon to Newcastle and Sunderland, there was a flourishing export trade in seaborne coal to London and the south by the seventeenth century.

As in Wharfedale and Swaledale, lead was an important product around Alston and in Weardale where the packhorses were called "carroways" (from carrier-galloways) and were fitted with muzzles to stop them grazing on the spoil heaps and so risking lead poisoning.

Last but by no means least in this Northern Region are the Southern Pennines, a high and bleak area largely composed of millstone grit and coal measures, with the high ground often covered by a thick layer of peat. Acting like a huge sponge, the peat ensures a year-round healthy flow of water into the many steep-sided valleys (cloughs). These streams provided the energy which powered early industrial activity.

This is no place for a history of the wool trade, but wool and its processing into cloth in the area around Rochdale, Halifax, Leeds, Wakefield and Huddersfield ensured intense packhorse activity in these "Industrial Pennines". The prevalence of the smallholder farms which made ends meet by cloth making at home (fulling, spinning, handloom weaving etc.) is well described by Defoe in *A Tour through the Whole Island of Great Britain*. He was travelling between Rochdale and Halifax in the 1720s. The change from these family-sized industries to the enormous valley-floor mills of the late nineteenth century has left an intriguing landscape and the area now exploits its past as a tourist attraction.

These hills can be bleak. Defoe again, in his classic account of crossing Blackstone Edge from Rochdale to the Calder Valley, even managed to experience a snowstorm in the middle of August.

Another interesting packhorse cargo has left its mark, even on present maps. The 1:25,000 OS map of the South Pennines (Outdoor Leisure 21) names two "Limers Gates", one in Grid square 89.23, the other in 00.30. They were both routes used by packhorses carrying lime from further north, used to sweeten the sour gritstone pastures around Rochdale and Halifax, as well as for use in making mortar.

To the west of the Pennines, the topography is altogether kinder and that intrepid horsewoman Celia Fiennes graphically describes her journey from Wigan to Lancaster in 1698, eating Clap bread and traversing many bridges (see Croston) on the way. She also mentions the crossroad guide posts which were erected following a statute of 1697.

CUMBRIA

Ashness Bridge (Group 3) 89/90: NY 270 196
Photographs of this bridge probably appear on more calendars than any other bridge in the country: stream and bridge in the foreground, woods and fellside in the middle distance, with the bulk of Skiddaw in the far distance. The small arch carries the narrow, winding road across Watendlath Beck from the Borrowdale valley to the hamlet of Watendlath.

It or a predecessor may have carried packhorse traffic from Watendlath to Keswick. In summer, it now carries a heavy burden of motorists.

Barbon (Group 1) 97: SD 614 818
Barbon is about two miles north of Kirkby Lonsdale, the bridge about a mile west of the village near Beckfoot Farm. A segmental arch of 24ft span crosses Barbon Beck alongside a ford. The bridge is narrow, only 27in between parapets which are 26in high and the parapet coping stones are connected by leaded-in iron staples. The roadway is grass-grown.

According to Jervoise, Barbon Bridge was described as "being ruinous" in 1725, so the present bridge (or repairs to the original) must be later. However, Hodge Bridge, in Barbon village is given

Barbon packhorse bridge near Kirkby Lonsdale, Cumbria

Ancient Monument status by the DOE, whereas the packhorse bridge is not. There is a question mark over which of these two bridges was ruinous in 1725. The packhorse bridge may have been on "Galloway Gate", a long distance packhorse and Drove road from Lancashire to Scotland which follows the River Lune hereabouts, though for it to have used Beckfoot Bridge en-route northwards from Kirkby Lonsdale it would have to have crossed the River Lune, either by the well-known Devil's Bridge at Kirkby Lonsdale (? fourteenth century), or by a ford. The crossing at Scar Brow (Grid Ref: SD 609 813) is a bridleway and a deep holloway descends to the Lune from the west.

From any route on the west side of the River Lune, Sedbergh is easily reached via Abbot Holme bridge which crosses the River Dee at Grid Ref: SD 649 909 and is late eighteenth century.

Birks Bridge (Group 2) 96: SD 234 993

This favourite bridge crosses the River Duddon which here flows clear, deep and green. The single segmental arch spans a natural gorge. The car park nearby and the beauty of the setting ensures a generous supply of people to sketch, take photographs, or just lean on the parapets.

The span is 12ft and the width between parapets, which are 40in high and have overflow drains built into their base, is 6ft 6in.

The bridge is on a track joining Cockley Beck to Grassguards, Stonythwaite and Wallabarrow.

Bleabeck Bridge (Group 3) 96: SD 188 919

This is one of the many small bridges which have at some time been widened, in this case probably on the downstream side. The original arch (upstream) is 7ft wide, the downstream widening increases this to 11ft overall and the join is obvious underneath the bridge.

I have chosen to include Bleabeck Bridge because of its original packhorse width and because of its location, crossing the Blea Beck just before it plunges steeply through woods to the River Duddon. The ruins of Frith Hall (originally a Medieval hunting lodge) a few hundred yards to the south, and the symmetrical little rock peak of Castle How a few hundred yards to the north, add to the interest of the site.

The bridge is of typical Lakeland "rustic" construction on a well-used farm track. The abutment footings and stream bed are encased in concrete, the span is 12ft and the width between 32in high parapets is 7ft 9in.

Boot (Group 2) 89: NY 177 012

This small bridge leads from the hamlet of Boot in Eskdale to the old Manorial Corn Mill. It crosses Whillan Beck in a span of about 27ft and the roadway is 6ft 6in wide between 38in parapets.

The corn mill is now a working museum with two sets of millstones driven by independent water wheels, both of which have been refurbished in recent years by apprentices from BNF at Sellafield.

The small exhibition inside the mill displays a padded wooden packsaddle on which were loaded two sacks of corn. The display also notes that there was once a regular weekly gang of 20 packhorses en-route through Boot from the west to cross Hard Knott and Wrynose on the way to Ambleside. From Boot they would probably cross the other Eskdale packhorse bridge - Doctor's Bridge (q.v.).

Burnbanks (Group 2) 90: NY 515 161

This bridge is named "Park Bridge" on Ordnance Survey maps and crosses Haweswater beck just downstream of the hamlet of Burnbanks and about three quarters of a mile downstream of the Haweswater dam.

The single segmental arch has a span of about 27ft and an overall width of 6ft 6in. There are no parapets but the bridge has modern wooden railings and the roadway appears to have its original surface. It is possible that the bridge is on an old route between Bampton Common and Shap (see also Rosgill) and/or from the lower end of the Haweswater valley before the level of the lake was raised by the Manchester Corporation Waterworks Department. It is also possible that it is simply a farm access bridge serving Thornthwaite Hall just to the north-west.

A short distance upstream, at Naddle Bridge, the old and disused arch adjacent to the present road bridge is also interesting.

Calva Hall (Group 1) 89: NY 059 265

DOE Listed Bridge. Cumbria County number 1.

This elegant bridge crosses the River Marron 4 or 5 miles east-south-east of Workington alongside Calva Hall farm. On top of the central upstream parapet stone is carved "HEAF 1697", presumed to be the building date and the builders initials. The bridge has a slightly asymmetrical segmental arch of dressed stone, spans 50ft and the width between parapets, is 44in. The parapets which are 11in high at the centre of the bridge, rising to 37in at the ends, flare markedly from the centre. The bridge is in good repair, the stonework of the western end having been recently pointed (1987). In a photograph of 1927, a wooden bridge is shown alongside which, between then and now, was "washed away".

The bridge is on an old packhorse route from Lamplugh (Grid Ref: NY 089 208) to Furness House (Grid Ref: NY 043 280). "Furness" is spelled "Furnace" on old maps suggesting an ore road and this is

An old bridge overshadowed by a massive arch of the Dent Head viaduct on the Carlisle-Settle railway, Cumbria

almost certainly the case because the remains of many disused iron mines can be found on the fells a mile or two south of Lamplugh. (See also Ullock.)

Crosby Ravensworth (Group 3) 91: NY 622 149
In a different location, this tiny bridge might be thought a packhorse bridge. However, a stream isolates the church of St Lawrence (parts of which may be as early as the twelfth century), from half of the village.

I suggest that the bridge is there to keep worshippers' feet dry.

Dent Head (Group 2) 98: SD 778 844
This strange one-arch bridge spans a feeder beck of the River Dee. The bridge is actually underneath one of the massive arches of the Dent Head railway viaduct on the Carlisle-Settle line.

There are no parapets and the original cobbled surface is covered by close-cropped turf. This bridge, like a footbridge described under Askrigg, North Yorks (q.v.) could be called "of inverse waisted shape", for the centre of the bridge is 15ft wide but the ends only 8 or 9ft wide. The span is about 30ft.

Sometimes referred to as a packhorse bridge, it is obviously wide enough for carts but since the present minor road which connects Dent village to the B6255 was not built until 1802, its status seems dependent upon its age. If post 1802, it was probably the road bridge until the building of the railway forced a road re-alignment. If pre 1802, then the bridge would form part of the old track connecting Dent with Horton-in-Ribblesdale via Gearstones (Grid Ref: SD 782 801) and High Birkwith (Grid Ref: SD 800 769).

Drigg (Holmes Bridge) (Group 1) 96: SD 076 987
DOE Listed Bridge. Cumbria County number 440.
Holmes Bridge crosses the River Irt, the outflow from Wastwater and here a substantial river. It is on the old packhorse road from Ravenglass (which once had a substantial market), via the ford across the River Mite at Grid Ref: SD 082 967 and Bell Hill, to Drigg and the north. The present length of bridleway, which connects Carleton Hall and Drigg church using the bridge, has disappeared under meadow apart from a few hundred yards at each end.

The single segmental arch has a high hump-back. The span is 39ft and the width 70in between parapets which are 31in high. The roadway has the original cobbled surface but the parapets are capped with newish-looking rounded copings of dressed red sandstone.

The word "Holme" is derived from Old Scandinavian and meant a piece of land partly surrounded by streams. It has come to mean "water meadow". (See also Bakewell.)

Eskdale (Doctor's Bridge) (Group 2) 89: NY 189 008

This lovely arch was originally a genuine packhorse bridge, but in 1734, the then resident of nearby Penny Hill, a surgeon called Edward Tyson had it widened so that he could cross in his pony and trap. The widening was completed without damaging the original arch and is obvious from underneath, the narrowest portion being upstream. The present dimensions are: span 36ft, width between parapets 10ft, parapets 30in high.

The bridge, which is in present use and well looked after, is at the junction of two packhorse routes from the west which joined to head east over Hard Knott and Wrynose towards Ambleside. One of these routes is from Ravenglass and can still be traced on the south side of the River Esk. The other, which uses Doctor's Bridge, is from Whitehaven via Skalderskew and Mitredale.

At the height of packhorse activity, Pennyhill was an Inn called "Pyet's Nest".

Far Easedale (Group 2) 90: NY 322 087

This crude arch crosses Far Easedale Beck just upstream of the confluence with the stream from Easedale Tarn. The span of 9ft is less than the overall width of 10ft 8in (there are no parapets). The shape is deformed, with holes through the roadway which is reinforced with lumps of concrete. In June 1989 the bridge surface, which is unused, had a flourishing growth of white stonecrop *(Sedum anglicum)*.

The bridge is known as the "Willie Goodwaller" bridge and legend claims that the original bridge was built without mortar over Willie's back as he stood in the stream acting as a template. This bridge was washed away in a flood and the present bridge is the replacement built by Willie's brother. It is thought that the bridge may be on the site of an even older bridge which served a fulling mill in Far Easedale.

The bridge is now on private farmland with no right of way and no path. It leads only from one field to another.

Furness Abbey (Group 2) 96: SD 224 715
DOE Listed Bridge. Cumbria County number 437.
Known as "Bow Bridge", this scheduled ancient monument is located a few hundred yards south of the impressive ruins of Furness Abbey on the northern outskirts of Barrow in Furness. Although the total span is only about 25ft, there are three semi-circular arches and the builders used the same red sandstone as for the Abbey. The overall width is just over 8ft. There are no parapets, though the English Heritage notice nearby suggests that there may have been stone parapets originally. If this were so, then Bow Bridge would most likely have been restricted to packhorse and foot traffic.

The bridge crosses Mill Beck which flows through and under the Abbey ruins, and once provided it with both water supply and drainage channel. The bridge was probably built in the fifteenth century, though the roadway has been recently repaired using concrete. That Bow Bridge is contemporary with the Abbey is confirmed by both its location and its appearance.

Gaisgill (Group 1) 91: NY 637 065
DOE Listed Bridge. Cumbria County number 449.
This is one of those packhorse bridges which, even when the Grid Reference is known in advance, still has to be searched for. About half a mile north of Gaisgill, a track leads east from the B6261, first to Fawcett Mill (a few hundred yards) and then Wain Gap (about one mile). The packhorse bridge is hidden in a clump of trees just upstream of Fawcett Mill and downstream of the bridge across Rais Beck which carries the track. The packhorse bridge is disused, but there is the faint trace of a hollow way leading to it from the west.

The almost semi-circular arch of 24ft span is without parapets and has an overall width of 66in. The roadway is of cobbles, now grass-grown.

The most likely purpose of the bridge was to serve Fawcett Mill,

The "hard to find" packhorse bridge near Gaisgill, Cumbria

but it may also have provided a crossing on a track from Ravenstonedale to join a major north-south route which headed north along the valley of the Lyvennet Beck and south through the Lune gorge.

Hartsop (Group 2) 90: NY 410 129

The bridge crosses Pasture Beck (from which it gets its usual name) at the upstream end of Hartsop village. As with other packhorse bridges in the Lake District, there are problems in tracing old routes from present evidence. Hartesope Hall (Grid Ref: NY 398 120) is marked on Saxton's map of Westmorland of 1577, but any track between Hall and village would have to cross either Kirkstone Beck or swampy ground around the head of Brothers Water. This problem also arises in connecting Hartsop with Ambleside via Scandale. It is more probably on the old packhorse route over Kirkstone Pass.

Sometimes referred to as Walker Bridge, the single arch spans 15ft. It has been widened in the past to its present width overall of 11ft. There are no parapets but the bridge has modern railings.

Hawk Bridge (near) (Group 3) 96: SD 236 918

This old bridge is worth recording, although I have found no evidence of commercial packhorse use. About half way between Hawk Bridge and the confluence of Appletreeworth Beck with the River Lickle, there is a group of ruined farm buildings and the bridge probably served for access from the farm to enclosed land on the north side of the beck.

Nearly semi-circular, the bridge has an 18ft span across the turbulent beck, here in a rocky gorge. The width is 8ft 6in between tumbledown parapet walls and the arch-stones are exposed for the central couple of yards of the roadway.

Widened at some time, the bridge is built in two distinct halves, the upstream half using mortar which drips miniature stalactites, evidence of significant age. The downstream half is probably even older since it is built without mortar. Whichever half was built first, it could have carried only people and animals and was probably widened to take farm carts.

High Sweden (Group 1) 90: NY 379 068

About 1 mile north of Ambleside, High Sweden Bridge crosses the Scandale Beck. The single, flattish and distorted arch which is vaguely semi-elliptical in shape, is similar to other "rustic" packhorse bridges in the Lake District in that it consists of arch-stones only. The span is about 14ft and the overall width about 68in. There are no parapets and the roadway has been repaired using concrete.

The purpose of this well-loved bridge is something of a mystery. The old packhorse road between Ambleside and Hartsop via Scandale Pass is still walkable, but it is on the same side (east) of the Scandale beck throughout. It may be that an alternative route from Ambleside via Nook End and Low Sweden bridge then re-crossed Scandale Beck to join the old route over the pass. It may also be that it was built simply to assist stock movement.

The bridge is now well-used by fell walkers as a means of getting from Scandale Lane on to the circuit of hills known as the Fairfield Horseshoe.

Horrace (Devil's Bridge) (Group 1) 96: SD 257 797

DOE Listed Bridge. Cumbria County number 450.

High Sweden - A fell walker's favourite.
Used to gain access to the "Fairfield Horseshoe", Cumbria

Located in the parish of Pennington near Horrace Farm about two miles west of Ulverston, the bridge is immediately alongside a minor road and crosses an unnamed beck which flows into Pennington reservoir. The grass-grown semi-circular arch spans 15ft and the width is 67in between kerbs which are about 10in high. The approach masonry of the bridge is almost parallel to the beck on both sides before the span proper turns to cross it. In a photograph of 1927, the minor road is shown as a ford, but the beck was culverted under the road during the 1939-45 war.

The bridge, though more recent than the Monastic period, is on an old ore road used by the monks of Furness Abbey which connected the iron ore mines in the Martin area to the south-west, with the bloomeries around Lowick (Grid Ref: SD 290 860). The ore was carried to the charcoal rather than vice versa.

One of the many bridges named after the Devil.

Lind End Bridge (Group 2) 96: SD 230 913

Carrying a public footpath across the River Lickle and located about half a mile upstream of the hamlet of Broughton Mills, this old bridge was probably built to carry packhorse loads of charcoal. In the woods covering the steep hillside to the north-west of the bridge, traces of old charcoal burning levels (pitsteads) can still be found. The public footpath through the woods is a holloway, quite steep and far too narrow for carts.

Within a few miles of the bridge, in the Duddon Valley near Ulpha (Grid Ref: SD 200 920) and on Torver Low Common (Grid Ref: SD 277 923), ancient bloomeries are marked on the OS 1:25,000 map, evidence that charcoal burning in this district has a long history.

Built in typical Lakeland "rustic" fashion, the bridge is without parapets, 8ft 6in wide overall and spans 15ft across a narrow gorge and deep pool. Wing walls level out the hump and at the centre, effectively form a secondary arch. The downstream keystone penetrates upwards into this secondary arch; upstream, two voussoirs at roughly 10 o'clock and 2 o'clock do the same.

There are signs that the bridge has been widened, the upstream part appearing more modern. If this is so, then perhaps the bridge's most recent employment was as a loading point for carts carrying charcoal away from the woods to the furnace at Backbarrow on the River Leven (less than ten miles away at Grid Ref: SD 358 849), which was using charcoal to produce high grade iron until 1920.

Monk's Bridge (Group 1) 89: NY 063 102

DOE Listed Bridge. Cumbria County number 441.

Probably named after the Monks of Calder Abbey, but also known as "High Wath bridge" (Wath is the Norse word for a ford), "Mattie Benn's bridge" (presumably the builder), "Hannah Benn Brig" (the builder's wife?) and "Roman bridge".

The bridge crosses the River Calder about 3 miles upstream of Calder Abbey at a remote spot and spans a spectacular, though small, natural gorge above a dark pool. The span is about 24ft and the total width (no parapets) about 4ft. The arch is pointed, asymmetrical, and looks on the point of collapse (and no doubt has looked so for a long time).

Monk's Bridge - "Mattie Benn's bridge" across the River Calder, Cumbria

Just downstream is a new wooden footbridge and a ford which was the crossing point of the old drove road from Gosforth to Cockermouth. At Thornholme, about 1 mile downstream there was once a bloomery. The bridge may have been used in the transport of iron ore for smelting.

Orton (Group 3) 91: NY 623 082
This bridge, built of arch-stones only and without parapets, joins a minor village road to what may once have been a common field. It is rudimentary and looks old, but if ever used by packhorses, I suggest it was only to carry produce (hay?) from the field.

Rosgill (Group 1) 90: NY 536 159
This single segmental arch crosses Swindale beck. The south-eastern end is built onto a natural rock outcrop over a pool on a right-angle bend in the river. The bridge has a 27ft span and is 41in wide between parapets which are 30in high, have been repaired and could be additions. The roadway surface looks original.

The bridge probably served a route between Shap and Bampton Common (see also Burnbanks) and is now well used by pedestrians walking the "Coast to Coast" route pioneered by Alfred Wainwright. There is an illustration in Wainwright's Guide.

The ruins of Shap Abbey (Premonstratensian) lie about one mile to the south-east.

Sadgill (Group 2) 90: NY 483 057
Sadgill, at the head of the narrow, six mile long Longsleddale, was a veritable "Crewe" for packhorse trains. Routes can still be followed down the dale to Kendal, over Stile End to Kentmere, north by Gatesgarth pass into Mardale, and north-east to Mosedale, thence to Shap or Penrith.

Early in the eighteenth century, complaints of delays to pack trains unable to ford the river when in spate, resulted in a petition for a bridge (1717). One was built and the original bridge (or a successor) has since been widened and now provides vehicle access to the farm. The widening is obvious underneath the bridge, the downstream section being about one third of the overall width.

36in parapets, topped by triangular copings, are now 9ft 6in

Rosgill - Walkers on Wainwright's Coast to Coast route
cross this bridge near Shap

apart. The segmental arch spans the River Sprint in about 30ft.

The rugged fells surrounding the dale head, the cluster of farm buildings, the stream and the bridge, form a typical Lake District picture.

Smaithwaite (Group 2) 90: NY 314 194

Smaithwaite bridge crosses the outlet stream from Thirlmere about half a mile downstream of the dam. The central stone arch has a span of about 15ft and the width, without parapets is 10ft. The arch is in the centre of the stream and is connected by rickety wooden footbridges to each bank.

There is a shallow weir just upstream of the bridge which gives the impression that the channel of the stream was perhaps widened and straightened as part of the Thirlmere reservoir works which were formally opened on 12 October 1894. If this is so, then the wooden footbridges must have been added to maintain the crossing and the right of way.

The bridge is on the route connecting Keswick and Grasmere via the east bank of Thirlmere and Dunmail Raise, but would also serve the old packhorse road over Sticks Pass from Glenridding via Stanah to Keswick. This route was used for transporting lead ore from the Glenridding mines to Keswick for smelting, an industry which started in the mid-seventeenth century.

Stainton (Group 1) 97: SD 525 859

DOE Listed Bridge. Cumbria County number 384.

It is difficult to establish that this bridge served other than local traffic. My speculation is that it was on a feeder route connecting south Lakeland with the main north-south packhorse and drove road which followed the Lune Gorge and was known as "Galloway Gate".

The small, segmental arch spans St Sunday's beck in the middle of Stainton village with a ford just upstream. The span is 18ft and the width between parapets, 36in. The parapets vary between 29 and 34in high and the bridge is built of roughly-dressed limestone.

Stang End (Slaters Bridge) (Group 1) 90: NY 312 030

This is a picturesque bridge at a picturesque spot in Little Langdale, crossing the River Brathay just downstream of Little Langdale Tarn. The whole is formed by a crude two-span clapper bridge joining an equally crude 'rustic' arch. The total span is about 75ft, the span of the arch about 5ft and its overall width 50in. The roadway surface is very rough. There are no parapets, but an iron handrail has been added. Every fourth or fifth arch-stone is untrimmed and projects above the roadway giving the bridge a lovely haphazard appearance.

One of the many packhorse bridges called Roman, the probability is that it was built to serve the slate quarries on the south side of the valley.

Stockley Bridge (Group 2) 89: NY 235 109

This was undoubtedly a packhorse bridge originally, crossing Grains Gill on the old packhorse road from Seathwaite in Borrowdale over Sty Head pass to Wasdale. "Sty" is a corruption via "Stee", of the Old Norse word *Stigi* meaning a ladder or a steep climb and the route was still being maintained for packhorse use in the early years of this century.

A favourite with Lake District fellwalkers.
Throstlegarth Bridge where the River Esk and Lingcove Beck join

The bridge was widened in 1887 and rebuilt in 1966 having been all but washed away in the disastrous Borrowdale floods of that year. The single arch presently spans 15ft and is 6ft 4in wide between parapets which are castellated and 21in high to the bottom of the castellations.

The roadway has been concreted and now carries a heavy load of fell-walkers. The resulting erosion to the causeway above the bridge heading for the top of the pass has been well repaired in the last few years.

Throstlegarth (Group 1) 89: NY 228 038
This magnificently sited bridge crosses the Lingcove Beck just as it joins the River Esk at the foot of the beautiful cascade of waterfalls and pools which guide the river down from the wild sanctuary of Upper Eskdale. It is a bridge which attracted early attention by the then HM Office of Works when, in 1928, they arranged for repairs to be made.

The farm of Brotherilkeld, where the track to the bridge leaves the metalled road was, in the fourteenth century, owned by the monks of Furness Abbey and was then called Butterilkelt. This association prompts suggestions that there was a Medieval bridge on the site, though without any evidence that the present bridge is Medieval. The bridge is also popularly called "Roman Bridge", perhaps because the Roman fort at Hard Knott is close by.

The bridge presently spans about 21ft and is 48in wide between kerbs which are 6in high. The centre of the span consists of arch-stones only and the roadway has been repaired using concrete. It is said to be on a route leading from Brotherilkeld via Ore Gap into the Langstrath valley where there was a smelting furnace or bloomery; a route used by iron miners. This is difficult to confirm from present evidence because the path between Brotherilkeld and Ore Gap is on the same side (east) of Lingcove Beck throughout. The bridge now provides access into Upper Eskdale and thence by Esk Hause to join the Sty Head path. It is sometimes called Lingcove Bridge.

Ullock (Group 1) 89: NY 074 240

This tiny bridge is at the back of the chapel near the centre of the village. It is behind the security fence of a woodyard which is known locally as "The American Sawmill" (because it is over the water).

Built from a single layer of rough arch-stones, the span is 12ft and the width 32in. There are nominal kerbs about 6in high. It is possibly on the same route as Calva Hall bridge (q.v.), ie. between Lamplugh and Furness House. The bridge crosses Black Beck which flows into the River Marron.

Walna Scar (Group 2) 97: SD 271 965

The bridge carries the well-known Walna Scar track over Torver Beck on its way between Coniston and the Duddon valley. The single arch is presently 96in wide between parapets and has a span of 12ft. The roadway is in good repair with cobbles set in concrete. The parapets are 18in high surmounted by iron railings.

Marked as "New Bridge" on a map of 1745 it has at some time been widened. Two distinct sections can be seen underneath the bridge, the upstream section having an overall width of 57in, the downstream 78in. Small stalactites hanging underneath the downstream section imply that this is the oldest.

Wasdale Head (Group 1) 89: NY 187 088
Probably the easiest packhorse bridge to reach in the Lake District, it crosses Mosedale beck just behind the Wasdale Head Hotel. The bridge has a 27ft span and is 48in wide overall. There are no parapets.

It is said to be on the route of the old corpse road from Eskdale via Burnmoor to Down-in-the-Dale and thence over Black Sail pass. If this is so, then the old route over Black Sail must have been on the opposite side of Mosedale Beck from where it is now. The more likely purpose is to serve a continuation of the well-known packhorse track over Sty Head pass which would have crossed the bridge and then continued down the north-west side of Mosedale Beck to the north bank of Wastwater before the bridge at Down-in-the-Dale (Grid Ref: NY 184 082) was built.

All authorities refer to this as a packhorse bridge but it is possible that it was built simply to help get sheep into and out of Mosedale.

Watendlath (Group 1) 89: NY 275 162
Perhaps the best-known and most photographed packhorse bridge in the whole of England, it crosses the beck at the outlet from Watendlath tarn. The span is about 22ft and the width, between 22in high parapets, is 41in. Early photographs show the bridge without parapets which are said to have been added between 1902 and 1907.

Of "rustic" construction and well-maintained, the bridge is on an old track and present bridleway between Rosthwaite in Borrowdale and Wythburn (Thirlmere). Another route probably crossed Ashness Bridge (q.v.), which is even more photogenic, lower down Watendlath beck on the way to Keswick. Both routes were used by pack ponies carrying wool.

It is my fancy that the rare and valuable "wad" or "plumbago" (graphite), which was mined from the hillside opposite Seathwaite for several hundreds years until the nineteenth century, was carried to Keswick over this bridge when the way through the "Jaws of Borrowdale" was difficult.

Water Yeat Bridge (Group 2) 96: SD 239 930
I first visited this bridge early in 1992 and because of its width (9ft

overall) and situation, decided that it had been built as a cart bridge. A year later, the fate of the bridge was in doubt. The County Council thought it unsafe and wanted to demolish it and build a stronger one. The resulting furore was reported in both local and national newspapers, where opinion consistently described it as a packhorse bridge. It seems I was wrong.

The bridge has been reprieved and a 3-ton weight limit imposed. It crosses the River Lickle about 2 miles upstream from Broughton Mills, by a shallow segmental arch which spans about 20ft. There are no parapets but the bridge is disfigured by bent and twisted iron railings. The centre of the span is composed of arch-stones only.

The (very) minor road of which the bridge is part, presently serves to connect three farms: Carter Ground, Jackson Ground and Stephenson Ground before joining the Coniston Road. These farms, with others, were formally established by Furness Abbey very early in the sixteenth century, which adds weight to the idea that the bridge is on a packhorse route connecting the ancient markets at Ravenglass and Hawkshead. There is an almost continuous line of lanes and bridle paths still connecting the two: from Ravenglass across the Esk by a ford at Waberthwaite, over the tops and across the Duddon at Ulpha, passing north of Stickle Pike, then via the three farms mentioned and across the bridge to join present tarmac roads to Coniston and Hawkshead.

However, the County Council date the bridge to the eighteenth century in which case it may have been used by pack trains carrying slate from quarries in the Dunnerdale Fells for shipment from Greenodd.

Wet Sleddale (Group 2) 90: NY 539 109

This crossing of the River Lowther is about half a mile upstream of the head of Wet Sleddale reservoir. The original bridge is said to be seventeenth century and was dismantled and re-erected by Manchester Corporation Waterworks Department at the request of the Friends of the Lake District to avoid being submerged when the reservoir was built. Early photographs show the bridge in its original location and without parapets.

A handsome and substantial bridge with a span of 30ft, it is 72 in wide between 33in high parapets. In its original position the bridge would have connected the farming area of Wet Sleddale on the south of the river, with the village of Shap. If genuinely seventeenth century, it may have had some connection with the old Deer Enclosures, remnants of which remain upstream.

Winster (Group 1) 97: SD 412 943

This most unusual bridge is about a quarter mile north of the village alongside the Windermere road (A5074). It is a single segmental arch of about 8ft span which crosses the River Winster near to its source. There is no filling above the arch-stones and there are no parapets. The overall width is about 72in. The bridge is immediately alongside the main road and the roadside wall is built on the line of the bridge up to and including the abutments. In a photograph of 1925 it seems as though the roadside wall was actually carried across the stream on the bridge, though for the span of the present road bridge, the wall has been replaced by railings.

The bridge was possibly on the pre-turnpike route between Ambleside and Kendal.

DURHAM

Deepdale (Group 1) 92: NZ 002 155

This superb segmental arch high above Deepdale Beck spans about 33ft. Although in my view an outstanding and early bridge (built in 1699), it seems to have escaped the notice of most authorities including Jervoise. He briefly mentions a "Deepdale Bridge" as being "in great decay" in 1605, but this presumably was either the bridge carrying the B6277 across Deepdale Beck just as it joins the River Tees, or the one carrying the minor road which connects Bowes and Cotherstone. I am grateful to Mr J.W.Fell of Cleasby for bringing the splendid bridge to my attention.

The bridge is 63in wide between 30in parapets and the track leading north-west to High Crag is a distinct "holloway". The route served by so substantial a bridge is puzzling. OS maps indicate a track joining North Thornberry and Nabb Farm (Nabb House on older maps) on the south of the beck, with High Crag and Low Crag on the north, before heading for Lartington.

The bridge was built by a Mr William Hutchinson to

Deepdale - A bridge in Durham.
Built by a man to commemorate being saved from drowning

commemorate being saved from drowning whilst crossing Deepdale Beck as a youth, and carries a stone inscribed:

> William Hutchinson of Melroo
> Esquire. Whose great charity was
> Most exemplary in all respects
> So likewise in the building this
> Bridge at Cragg, the place of his
> Most happy nativity. which was
> Built in August 1699
> Edward Addison fecit

Egglestone Abbey (Group 1) 92: NZ 062 152

DOE Listed Bridge. Durham County number 129.

This seventeenth-century bridge crosses Thorsgill Beck in the shadow of Egglestone Abbey, upstream of and immediately alongside the present road bridge. The almost semi-circular arch has a span of 27ft and is 66in wide between 18in parapets. The roadway surface looks to be of the original cobbles, but is now grass-grown.

The bridge served an old road on the south side of the River Tees between Barnard Castle and Rokeby and is mentioned in a book by Sir Richard Colt Hoare describing his journeys through Wales and England undertaken between 1793 and 1810.

Headlam (Group 1) 92: NZ 179 189
DOE Listed Bridge. Durham County number 18.
Just off the village green this small semi-circular-arched bridge crosses Dynance Beck which at this point is a slow-moving stream meandering its way through a swampy hollow. A length of stone-built causeway leads to the bridge on either side. The span is 9ft and the width 70in between rudimentary 12in parapets or kerbs which appear to have been added and have stapled copings. The roadway was grass-grown and the structure not very well maintained in 1988. By 1993, the bridge had become so badly overgrown as to be almost hidden. Pevsner assigns a date of seventeenth or eighteenth century.

It is difficult to propose a packhorse route that this bridge may have served.

Ketton Hall (Group 1) 93: NZ 302 192
DOE Listed Bridge. Durham County number 56.
Hereabouts is a very old bridge site; a Ketton Bridge is mentioned in Durham Abbey papers in 1274. The existing packhorse bridge crossed the River Skerne which was re-routed in the second half of the nineteenth century, obviously later than the bridge was built because the river is presently 100yds away and the bridge isolated in the middle of a field. The single arch has a span of 24ft and is 57in wide between parapets which are 25in high. The cobbled roadway looks original. Sunk into the ground nearby is a cast iron notice which reads "Ketton Road Ends Here," probably a Parish boundary marker defining the limit of responsibility for maintenance.

Presently (1990) the land surrounding the bridge is intensively farmed and it seems that in 1988, the green lane which was known as Salters Lane and of which the bridge was part, was destroyed. Any Salters Lane has had an obvious use; the salt is most likely to have come from evaporating pans on the coast.

The cast iron notice close by Ketton Hall packhorse bridge in County Durham

Thwaite (Group 3) 92: NZ 039 112 and 113.

In the hamlet of Thwaite is a small chapel (anonymous, but presumed Methodist). From the minor road between Thwaite and Scargill, a bridleway leads to the chapel across an unusual stone bridge with a small rectangular opening for the un-named stream. The bridge has a roadway 57in wide running between parapets for about 30ft although the water passage is only 33in wide by 54in deep. The parapets are 30in high with rounded copings.

On the other side of the chapel, a wooden footbridge spans Thwaite Beck a few hundred yards before it joins the River Greta. This footbridge is founded on stone abutments which appear to have provided the springing for an arch. Also, a cobbled roadway descends from the chapel to what looks like a ruined ford.

The bridleway status of the track and the ruins of bridge and ford across Thwaite Beck suggests an originally more substantial use for these two crossings than simple access to the chapel. They are interesting constructions whether or not they formed part of a packhorse route.

West Hope (Group 1) 92: NZ 032 093

The hamlet of West Hope is about half a mile east of the Stang Road, which leads from Arkengarthdale northwards over Hope Moor to join the A66. More significantly, West Hope is on an almost continuous line of lanes and bridleways between Bowes and Richmond, parts of which are known as "The Badger Way" (see Glossary).

Crossing just after the confluence of Waitgill Beck and Hill Beck is a tiny bridge in the shape of a shallow segment. The bridge spans about 12ft, is 42in wide, has no parapets and is formed of archstones only with a roadway of flat slabs. Erected on top of it is a modern wooden footbridge complete with handrails.

The position of this bridge on an old route so important as to be named, persuades me that, diminutive though it is, it probably had genuine packhorse use.

GREATER MANCHESTER

Prestolee (Group 1) 109: SD 751 062

Both the site and the construction of this bridge are unusual. It crosses the River Irwell just upstream of the confluence with the River Croal and is about half a mile north-west of Prestolee church. Three bridges span the river within a few tens of yards. Downstream is the packhorse bridge, then comes a modern concrete structure in the form of an arch above the roadway (like Sydney Harbour Bridge). In this case the roadway is a closed concrete box, presumably carrying a utility such as water or gas. Upstream of this is a magnificent four-arch aqueduct carrying the defunct Manchester, Bolton and Bury Canal. Although much of the canal is now dry, the aqueduct and about a mile of canal towards Manchester are still flooded.

Prestolee - The only packhorse bridge in Greater Manchester

The bridge site is unusual in that it lies in a pocket of rural land lost in nineteenth-century industrial suburbs. Bolton and Bury enclose the north and Manchester the south; Farnworth and Whitefield block west and east respectively. Adding to the rural illusion is a group of dilapidated farm buildings between village and bridge, one of which is a cruck-framed barn.

Prestolee packhorse bridge spans the Irwell in five semi-circular arches springing from high piers (the river is in a shallow gorge) which have triangular cutwaters both upstream and downstream. The total span is about 150ft and the concreted roadway is 60in wide between 43in parapets. On the vertical inward face of the upstream parapet at about mid span is carved: 'CC Prestolee Bridge'.

According to Jervoise, the bridge was built between 1781 and 1803, so it was contemporary with the canal which was first in use in the 1790s.

The bridge probably served a route connecting Manchester and Bolton, but was perhaps also used, as was the canal, for the transport of coal, mined and used locally.

LANCASHIRE

Bleasdale (Group 1) 102: SD 566 458
DOE Listed Bridge. Lancashire County number 77.
This attractive listed bridge spans the River Brook near Brooks Farm, Bleasdale. A wide 'V' of low walls on the south bank leads into the bridge from a small patch of bluebell wood; the other end joins the farm with a series of small drystone wall enclosures. The farm buildings, walls and bridge together make a satisfying and pleasant grouping.

The bridge is small, spanning about 13ft with a width of 33in between 14in high kerbs. The roadway is flagged.

One possible route using the bridge is from Lancaster in the north-west, skirting south of the high ground of Grizedale Fell, Bleasdale Moors and Blindhurst Fell, heading via Chipping towards Clitheroe. The walled enclosures (for stock control?) suggest that the bridge was built as much for farm use as for long-distance packhorse trains.

Catlow Bottoms (Group 1) 103: SD 885 363
Located in the aptly named Catlow Bottoms and spanning Catlow Brook, this packhorse bridge is one of the few with castellated parapets. The arch is semi-circular and spans about 10ft, with a width between parapets of 52in. The parapet castellations are 6in square on a 6in pitch and the height to their base is 33in. The roadway is cobbled with the centre of the arch protected by a thin skin of concrete.

The bridge served two packhorse routes; one between Colne and Burnley, the other, one of the "Limers Gates" between Clitheroe and Halifax. Alcomden Bridge and Lumb Bridge, Crimsworth Dean (q.v.) are both on this latter route.

Croston (Town Bridge) (Group 2) 108: SD 489 185
DOE Listed Bridge. Lancashire County number 58.
Celia Fiennes, exploring much of England on horseback in the 1690s, describes her journey from Wigan to Preston in 1698 as "passing by many very large arches that were only single ones". She continues "they are but narrow bridges for foote or horse ... I passed by at least half a dozen of these high single arches ... over their

This bridge in Croston village, Lancashire, may have been used by the seventeenth-century traveller Celia Fiennes

greatest rivers".

The only significant streams barring progress between Wigan and Preston are the Rivers Douglas, Yarrow and Lostock and Town Bridge crosses the River Yarrow in the middle of Croston Village. The central parapet stone on the downstream side is dated 1682, so it is my fancy that this is one of the "many very large arches" traversed by Miss Fiennes in 1698.

The single, segmental sandstone arch spans 39ft and is about 8ft 6in wide between parapets which vary from 16in high at the ends, to 37in in the centre. In use by motor traffic subject to a weight and width limit, the cobbled roadway is named "The Hillocks". The copings are stapled across the top and some of them also vertically to the parapet stones beneath. Shaped iron straps, just above the arch stones on each side of the bridge and following the curve of the central two thirds of the segment, are connected by four big through bolts. The arch itself is in good repair, the parapets less so. A disfiguring water pipe is clamped to the bridge on the upstream side.

Higherford Old Bridge (Group 2) 103: SD 863 401

DOE Listed Bridge. Lancashire County number 51.

This beautiful and substantial late sixteenth-century bridge is known variously as "Th'Owd Brig" and "Roman Bridge". It crosses Pendle Water in a high segmental arch which spans more than 40ft. It is located in a quiet corner of Higherford which still has the character of an old village; one cottage is dated 1755.

Parapets with square copings which are stapled together, rise to a shallow point and are 40in high. The width between them is 8ft 6in and the roadway is cobbled. As befits a listed monument, the bridge is in very good repair.

In *The Bridges of Lancashire and Yorkshire,* Margaret Slack suggests that the bridge was used by packhorse trains carrying coal from near Gisburn and by "Lime Girls from Lothersdale bringing lime into the district".

Hodder - Old Lower Bridge (Group 2) 103: SD 704 391

DOE Listed Bridge. Lancashire County number 20.

The Hodder is a substantial river which drains the southern slopes of the Bowland Fells. About a mile upstream from its confluence with the River Ribble near Great Mitton, the Hodder is crossed by the B6243 and just downstream of the modern road bridge is the old and disused packhorse bridge.

Built at the instigation of Sir Richard Sherburn in 1561, the bridge is still a substantial structure, but can only be reached by climbing a high chestnut-paling fence, a challenge the writer avoided.

There are three segmental arches, the largest one in the centre now comprising arch-stones only. The two outer arches still contain some rubble filling and there are traces of what could have been retaining or parapet walls on the western arch. Each pier has a full-height triangular cutwater upstream.

According to Jervoise, the total span is 126ft and the overall width approximately 7ft.

Wycoller (Group 1) 103: SD 932 392

DOE Listed Bridge. Lancashire County number 109.

Located in Wycoller Country Park established by Lancashire County Council, this interesting thirteenth- or fourteenth-century bridge is

Paddling time in Wycoller Country Park, Lancashire

surrounded by many remains with historical significance. Close by are: an old clapper bridge, a clam bridge the park leaflet claims to be of Iron Age date (which seems doubtful), a superb aisled barn, thirteenth-century vaccary walls formed of vertical sandstone flags, and the ruins of Wycoller Hall which by repute is the model for Ferndean Manor in Charlotte Bronte's novel *Jane Eyre*.

The bridge is formed of two arches (one of which is deformed) with a total span of about 30ft. The width is 26in and there are rudimentary kerbs 10in high. The central pier has a small upstream cutwater and the roadway is cobbled and very rough. The bridge, which spans Wycoller Beck, is on an old packhorse route between Colne and Keighley and is sometimes known as "Sally's Bridge".

NORTHUMBERLAND

Ovingham (Group 1) 88: NZ 083 636
The bridge crosses the Whittle Burn in the middle of the village.

Beaumont Clough, near Hebden Bridge, Yorkshire *(A.Menarry)*
Eastwood Bridge, Calderdale *(A.Menarry)*

Foster Lane (or Hollins Lane), Hebden Bridge *(A.Menarry)*
Ragby Bridge over Ramsden Clough, near Todmorden *(A.Menarry)*

There are two segmental arches with triangular cutwaters both up and downstream on the central pier. The total span is 42ft, and the width between parapets 59in. The parapets, which appear to be additions, are 30in high with stapled copings.

It seems that "Ovingham bridge fell down in a great fire in 1697" so the present bridge is probably a repair or rebuild from 1698. The bridge serves a route along the north bank of the River Tyne between Wylam to the east and Ovington to the west. Although nearly always referred to as a packhorse bridge, Jervoise describes it as a footbridge. The engraver Thomas Bewick is buried in Ovingham church.

NORTH YORKSHIRE

Aldbrough (Group 1) 93: NZ 202 114
DOE Listed Bridge. North Yorks County number 415.
The bridge is in Aldbrough village about 4 miles south-west of Darlington and crosses Aldbrough Beck by three pointed arches which are of different sizes. The total span is about 48ft, the width between parapets 53in and although the present parapets are 44in high, there are indications of original kerbs only 6in high. The construction is rough, with one abutment built into the foundations of a cottage, or more likely, a cottage founded on the abutment. There are two upstream cutwaters of triangular section; the two downstream are rounded. The roadway is of recent concrete.

The Roman road Dere Street passes less than one mile east of Aldbrough on an almost due north-south alignment and crosses the River Tees at Piercebridge, the site of the original Roman bridge. The present Piercebridge dates from the early sixteenth century, though there has been rebuilding and widening since. About two miles south is Lucy Cross from which a minor road heads south-west towards Aldbrough. Piercebridge, Lucy Cross and Aldbrough lie roughly on the line of an important packhorse road connecting Tyneside and Lancaster, proceeding from Aldbrough towards Lancaster via Marske, Hawes and Ingleton.

Askrigg (Group 2) 98: SD 935 910
DOE Listed Bridge. North Yorks County number 81.

Known as "Bow Bridge", this crossing of Grange Gill Beck (or Sargill Beck) is alongside the minor road which runs parallel to the Ure on the north bank of the river. There are no parapets and the roadway (unused) is grass-grown, but at 24ft wide, the usual description as a packhorse bridge seems fanciful. The semi-circular arch has four ribs (the outer ones chamfered), a feature which confirms its probable Medieval origin. It is thought to be early thirteenth century, built by the Monks of Jervaulx Abbey. It has been widened on the downstream side.

About 200yds downstream, near the old railway bridge there is another interesting bridge only 10in wide at the entrance and 23in in the middle; 26in parapets are topped by iron railings. The 10in width plus a wooden gate at each end effectively stops much smaller animals than packhorses. It was evidently built as a footbridge.

Barton (Group 3) 93: NZ 231 086
This three-arch bridge alongside a well-used ford in the village is of interesting construction. The two outer arches approximate to semi-ellipses, the central arch is segmental and there is a square overflow opening at the south-eastern end. The total span is about 120ft and the overall width (no parapets), about 8ft. The arch-rings are brick and the filling, rubble. There are substantial white-painted wooden railings in lieu of parapets.

I have found no hint of packhorse use, but age, narrowness and the unusual construction merit the bridge's inclusion.

Baysdale Abbey (Group 3) 94: NZ 621 068
This pointed arch with four square ribs, is undoubtedly medieval and now leads only to Baysdale Abbey farm, the Abbey itself having gone. The span of about 9ft is less than the width between parapets of 9ft 6in. There are modern kerbs and 20in parapets with square copings.

Although wide enough for carts and now carrying motor traffic, its age, remote location and Monastic use prompts inclusion.

Birstwith (Group 1) 99: SE 236 603
DOE Listed Bridge. North Yorks County number 138.

Called locally "New Bridge", or "Haxby Bridge", Birstwith is one of the lovely long-span high arches in the Yorkshire Dales.

The span is about 66ft and the width 6ft 6in between 36in parapets topped by massive square copings. At each end of the bridge is a pair of tapered round pillars just within the parapets, with the presumed function of excluding carts, since they limit the effective width to 49in at roadway level and 62in at parapet level.

A bridge across the River Nidd has been known at this site since the end of the sixteenth century and the present bridge was built in 1822. This is later than my self-imposed rule, but for this bridge I am happy to break the rule.

It is difficult to propose a packhorse route for so late a bridge. My speculation is that it connected the southern side of the Nidd valley with one of the more northerly market towns such as Kirkby Malzeard.

Boltby (Group 1) 100: SE 491 866

This tiny, almost semi-circular arch crosses Gurtof Beck in the middle of Boltby village. It now forms a part of the footpath along the main street and is no doubt still useful when the road floods.

Although its stature as a packhorse bridge is dubious, I have included it because of its position on the route from Hawnby in Ryedale, to Thirsk. This line of now minor roads crosses the "Hambledon Drove" just over one mile to the north-east of Boltby.

The bridge has a span of about 9ft and is 46in wide between parapets which rise to a shallow point at the centre where they are 36in high.

Many parapet coping stones on old packhorse bridges are fastened together by leaded-in iron "staples", presumably to prevent their being pushed easily into the stream. At Boltby, a single 2in by $^1/_2$in iron strap the full length of each parapet is bolted to the copings. A jack hammer would be needed to disturb them.

Caldbergh (Ulla Bridge) (Group 2) 99: SE 091 851

In some ways this bridge is included under false pretences, but its site and construction details are unusual, hence its description.

The unnumbered minor road on the south bank of the River Cover crosses Caldbergh Gill by a skew bridge built in 1933, just upstream of which is the older and now disused Ulla Bridge. The Gill at this point is a vertically-walled canyon worn into the limestone which would be very difficult for human beings to cross without a bridge. For horses quite impossible. There may be easier crossings upstream and/or downstream, but the bridge has been there since packhorse days as is obvious from its construction.

The grass-grown roadway spans about 15ft and is about 13ft wide with remnants of dry-stone wall on both sides. For those unworried by muddy shoes, it is the underside of the arch which is interesting. Seen from the water-worn limestone underneath, there are three quite distinct arches.

The downstream arch is 72in wide, is the crudest construction of the three and grows the longest stalactites, so it is probably the oldest. The middle arch is 50in wide and the upstream 56in.

I speculate, without documentary support, that the bridge started life as a packhorse bridge of 6ft overall width and was widened upstream in two distinct stages before being superseded.

Clapham (Group 1) 98: SD 745 692

This segmental arch of 21ft span crosses Clapham Beck in the middle of the village. The width is 41in between parapets which are 38in high with rounded coping stones. The line of the bridge has a slight wriggle.

Although important for local village use, the bridge may have seen longer distance traffic. Clapham lies on the line of an old road from Lancaster and Kendal/Kirkby Lonsdale in the west, to Settle via Austwick and Feizor. Many routes with known packhorse associations radiate from Settle and these are well-described in *Roads and Trackways of the Yorkshire Dales*, by G.N.Wright.

Of this bridge Jervoise writes, "The masonry is very rough and it is quite impossible to judge its age."

Croft (Group 1) 93: NZ 281 100

Croft-on-Tees, about three miles south of Darlington, is one of the historic crossing points of the Tees, the present bridge probably dating from the fifteenth century. Immediately upstream, the Tees is joined by Clow Beck and at the end of the lane leading to Monk End Farm, about half a mile west of Croft bridge, there is an unusual packhorse bridge with two arches which are so far apart they almost form two bridges. To add to the oddness, the two arches are on slightly different lines (a bent bridge).

The total span is about 40ft and the cobbled roadway is 48in wide between 20in parapets which are stapled together. There is a recently built (*circa* 1965) ford used by farm traffic, upstream. The arches are segmental, of unequal size and the pier carries a rudimentary cutwater upstream.

It is thought that the larger arch (the western) formed the original bridge and is contemporaneous with the main Tees bridge at Croft. What is now the midstream pier was originally the buttress for this larger arch, perhaps with a causeway leading eastwards. The smaller arch was constructed at a later date.

The purpose of the bridge was probably to assist communities south of the Tees and west of Croft (Stapleton, Cleasby, Manfield) cross Clow Beck and gain access to Darlington via the Tees Bridge at Croft. *Note* Until the building of Blackwell Bridge in 1832, Croft was the only crossing of the Tees into Darlington from the south.

Danby (Group 2) 94: NZ 719 078
DOE Listed Bridge. North Yorks County number 48.
Danby, Glaisdale and Westerdale are three magnificent old packhorse bridges which cross the River Esk. They are all described as Group 2 bridges, not because they are inferior in any way, but because they all exceed 6ft in width. This bridge, which was formerly called Danby Castle Bridge, now goes by the name of Duck Bridge because it was repaired by one George Duck early in the eighteenth century (Sessions papers of 1717 record payment of £10 7s 6d to help the cost of repair).

The bridge bears the coat of arms of the Neville family, who followed the Latimers at Danby Castle in abut 1380. Pevsner dates the bridge to 1386. The hump is high, the span abut 45ft and the width 75in. The parapets are 43in high with stapled copings and bear scars and scratches from passing cars which use the bridge. There are indications of two lower levels of parapet.

Originally providing access to Danby Castle, the bridge was later the river crossing of a probable north-south route using the spur of high ground called Danby Rigg, leading to the southwards-opening Rosedale, site of Rosedale Abbey (a Cistercian Nunnery founded in 1190).

Dob Park (Group 1) 104: SE 196 509

This high segmental arch across the River Washburn between Swinsty reservoir and Lindley Wood reservoir, is on a route northwards from Otley in the Wharfe valley towards Summer Bridge in Nidderdale.

From at least the sixteenth century this has been the site of a bridge, the Parish of Fewston then being responsible. The present bridge probably dates from 1738, when a "ruinous and irreparable" bridge was rebuilt for £50 funded by the Township of Clifton with Norwood.

One of the many graceful bridges in the Yorkshire Dales, Dob Park is substantially built in a gritty sandstone. The span is about 48ft and the width 55in between 24in parapets which are topped by square copings stapled together. The roadway is cobbled and a ford downstream is used by farm traffic.

Dousgill Farm (Group 3) 92: NZ 093 081

I have found no evidence that this is other than a bridge providing access between farm and fields, but it is small, looks old and is in a remote spot, hence its inclusion.

A causeway 30ft long leads from the farmyard to the bridge and another of 12ft from the bridge (which spans about 5ft) towards the fields. The upstream parapet is continued as a field boundary wall; the downstream is in ruins; the width between, just over 7ft.

The modern traffic of the farm uses a ford just downstream.

Glaisdale (Group 2) 94: NZ 784 055

DOE Listed Bridge. North Yorks County number 56.

Normally called Beggars Bridge, this lovely high arch crosses the River Esk alongside modern road and railway bridges. The span is 54ft and the width 6ft 6in between 32in parapets. The arch is ribbed, has arch rings in two orders and is dated 1619. It bears the coat of arms of Thomas Ferris of Hull and was known formerly as Ferris's Bridge.

The legend attached to the building of this bridge is worth re-telling. It is said that Ferris, a poor farmer's son from nearby Egton,

Glaisdale - This North Yorkshire bridge across the River Esk figures in a romantic legend

fell in love with Agnes Richardson the daughter of the Squire of Glaisdale. He used to visit using stepping stones which were then the only way across, but because of his poverty he was an unwelcome suitor. To make his fortune, he signed on with a ship which was joining one of Sir Francis Drake's voyages. On the night before sailing he went to say goodbye, but the river was in flood and he could not cross. He swore to build a bridge at the spot and having returned with a fortune, did so. He also married Agnes and became Mayor of Hull.

Apart from its usefulness to courting couples, the bridge provides access from Glaisdale with its flagged causeway for packhorses, to the almost direct high-level road to Whitby by way of Egton and Aislaby.

Hampsthwaite (Group 1) 104: SE 260 587
DOE Listed Bridge. North Yorks County number 331.
The bridge is located a few yards down a lane which continues the line of the footpath which bisects the triangular village green. Since

the DOE name the bridge "Cockhill Packhorse Bridge", I assume that it crosses Cockhill Beck, unnamed on the 1:50,000 OS map.

The bridge is small, a semi-circular arch of about 12ft span and 54in wide between 26in parapets with stapled square copings. A line of field paths and minor roads lead south-eastwards from Hampsthwaite and may have crossed Oak Beck by Knox packhorse bridge (q.v.) to reach Harrogate.

A prohibition against cycling and a high hump is an irresistible combination to small boys trying to get both wheels airborne.

The main bridge which crosses the River Nidd at Hampsthwaite is also interesting. According to Jervoise, the Sessions in 1640 decided to "new build in stone" at a cost of £400 "owing to the scarcity of wood".

Harrogate (Group 2) 104: SE 285 555

Oak Beck, running through a rather steep-sided and waterlogged valley, forms the boundary of the north-western quarter of Harrogate. The valley bottom is well-wooded and Irongate Bridge (its local name) comes as a surprise so near the centre of a large town (the bridge is only a mile or so from Harrogate railway station).

The semi-circular arch has a high hump and spans about 20ft. There are no parapets and the overall width is 6ft 4in. The roadway is cobbled and a few arch-stones stick up above roadway level. 8 or 9yds. of cobbled causeway lead to each end of the bridge at an angle, so that in plan bridge and causeways form a crude "Z" shape.

Jervoise refers to Irongate as a packhorse bridge and certainly the high hump and angled causeways would cause problems for carts. There is an old road between Beamsley, to the west, and Harrogate, some stretches of which are called "Badger Gate". Irongate Bridge, at the Harrogate end of this route, fits too neatly to be a coincidence. This, plus the nature of the valley of Oak Beck, convinces me that Jervoise was correct.

Hubberholme (Group 1) 98: SD 935 788

This crude old bridge spans Crook Gill just before it joins Cray Gill which in turn joins the River Wharfe about half a mile downstream. The bridge is about half a mile north-east of Hubberholme church.

The bridge shape is a very shallow segment spanning about 13ft

6in and the width (no parapets), is 72in. The downstream keystone is a large and crudely-shaped piece of limestone which presents a face 36in deep by 10in wide and which penetrates 20in into the arch. Other arch-stones are as crudely-shaped though smaller and some of them stick up above the roadway to form an intermittent kerb.

The roadway surface is roughly cobbled and is built up for 30ft or so from either end of the bridge. Although of extremely rough appearance, the bridge is in good repair, pointing on the underside appearing recent (Oct 1991).

The bridge is on a track connecting Bishopdale and Wharfedale.

Ivelet (Group 2) 98: SD 933 978
Often referred to as "The Queen of Swaledale Bridges" this superb arch soars gracefully across the Swale in a high hump-back. Built in the 1690s and not subsequently widened, it carries a very minor road with a weight limit of 2 Tons and is labelled YNR 349. The span is at least 60ft and the width about 8ft 4in between 37in parapets.

Although often referred to as a packhorse bridge, it is wide enough to accommodate farm carts and now carries motor traffic.

Apart from the bridge's obvious use in joining the north and south banks of the River Swale, it is also at the northern end of a track from Wensleydale over Askrigg Common.

Knox (Group 1) 104: SE 295 578
DOE Listed Bridge. North Yorks County number 297.
With a ford just downstream, Knox (or Spruisty) packhorse bridge crosses Oak Beck about a mile before its junction with the River Nidd.

The bridge presents an oddly formal appearance, each entry being guarded by two quasi-ornamental pillars. The width is 49in between substantial parapets which are 30in high with stapled copings. The segmental arch is nearly a semi-circle and spans about 18ft. The roadway is of modern paving slabs and the wing walls on each side are pierced by square overflow ducts.

Apart from possible use in connection with a nearby disused mill, the bridge would provide a crossing of Oak Beck on any route northwards from Harrogate towards Ripon, or westward into the Nidd valley. (See also Hampsthwaite.)

The inscribed stone on Ling Gill Bridge, North Yorkshire

Ling Gill (Group 2) 98: SD 803 789

Ling Gill is one of the classic, remote packhorse bridges which are often described. Located on the important Settle - Horton-in-Ribblesdale - Hawes packhorse road, some stretches of which are Roman, it now carries a heavy traffic of walkers on the Pennine Way long distance footpath.

Wide (16ft between 40in parapets with rounded copings) it nevertheless is a true packhorse bridge and shows no signs of having been widened. The arch is an exceptionally shallow segment spanning about 20ft and springing from about 4ft above stream level. The stream is called Cam Beck upstream of the bridge, Ling Gill Beck downstream and reverts to Cam Beck before it joins the River Ribble about 2 miles downstream.

The upstream parapet carries a stone inscribed:

ANNO 1765
THIS BRIDGE
WAS REPAIR

ED AT THE
CHARGE OF
THE WHOLE W
EST RIDEING

The 'Ns' in 'ANNO' are reversed and the date is difficult to decipher between 1765 and 1768.

Linton in Wharfedale (Group 1) 98: SD 997 627

DOE Listed Bridge. North Yorks County number 313.

The bridge crosses Eller Beck in the middle of the village just downstream of a clapper bridge which has been moved upstream from its original position. Although generally referred to as a packhorse bridge, the legend of its building hints at a footbridge. It seems that two ladies who lived in the village in the sixteenth century were unable to get to church when the clapper bridge (which pre-dates the packhorse bridge), was under water during times of flood. Failing to get financial help from local farmers to build a bridge, the ladies did so at their own expense, and made it too narrow for farm carts to use. The bridge was repaired in the late seventeenth century at the expense of a lady called Elizabeth Redmayne and is known as Redmayne packhorse bridge. There is one main arch with a side-arch overflow. The total span is 60ft and the width between parapets 42in. The parapets are 36in high.

Marske (Pillimire Bridge) (Group 2) 92: NZ 100 007

This substantial bridge across Marske Beck is another of the puzzles afflicting the writer searching for facts. None of the well-known guides (including Jervoise) refer to it. Although 8ft 6in wide between 36in parapets and thus wide enough for farm carts (and now used by farm stock), it presently carries only a footpath, not a bridleway.

There are two segmental arches spanning a total of about 54ft. The north-eastern arch is dry, across land which is several feet above normal river level.

One clue as to the purpose of a possible earlier crossing at this point is an almost direct line of paths between Marrick Priory (to the south-west) and across the river towards Marske Moor. The Priory was a foundation of Cistercian and Benedictine Nuns, so sheep and their grazing grounds would be important Priory assets in the

Middle Ages.

The present Pillimire Bridge however, is thought to be late eighteenth century and at that date, the most likely packhorse cargo to use the bridge would be lead ore.

Marske Bridge, only a few hundred yards downstream, probably dates from the fifteenth century, so the purpose of Pillimire Bridge, though it is said to serve an old packhorse route, remains a puzzle.

Immediately upstream on the south-west bank is a derelict overshot waterwheel, apparently an abandoned attempt in the late nineteenth century to drive an electric generator to supply nearby Skelton Hall.

Masham (Swinton Park) (Group 2) 99: SE 197 800

Across a very deep gorge which cuts through the boundary of Swinton Park is a Victorian viaduct built in 1832 and carrying a minor road. The viaduct, called Quarry Gill Bridge is large, has battlemented parapets and spans the gorge at high level. Underneath this viaduct on the downstream side is a small arch which spans the stream at low level and which has a grass-grown surface. There is also a similar looking structure at the same low level upstream. Access to this low level is both difficult and dangerous and the writer failed to make it. There is no evidence of any trackway from present road level to stream level.

Pevsner (North Riding) quotes - "Deep down below [is] the old packhorse bridge now covered in grass and bracken." It seems somehow unlikely. The appearance from above is of a low level arch built to support the piers of the high level bridge.

Newsham Lodge (Group 3) 92: NZ 101 109

Newsham Lodge (a farm) is late eighteenth century and the bridge probably of the same age. It is a "bridle bridge" in that beyond the farm northwards, a bridle path continues up to the A66. In the past, the track continued to Barnard Castle.

A 30ft long causeway leads to the bridge from either end and the bridge itself spans about 12ft. The width is 11ft between 22in parapets with square copings. The roadway surface looks to be original.

Ings Bridge, just south of Pickering town centre

Pickering (Group 1) 100: SE 791 820

The bridge, known locally as Ings Bridge, crosses Pickering Beck about one mile south of the town. The segmental arch spans 25ft and the grass-grown roadway is 51in wide between 36in parapets whose copings are joined by iron staples. "Ing" is derived from the Old Norse "Eng" meaning meadow or pasture, an apt description for the area either side of the bridge.

Ings Lane/Landales Lane and Haygate Lane/East Ings Lane follow either side of Pickering Beck southwards from the town and provide access to a network of lanes and fields. Ings bridge facilitates a passage between these two networks and was probably built for this local traffic. Pickering Low Mill is just upstream on the west side, so the bridge would also provide access to it from fields on the east.

Much land in the Vale of Pickering goes by the name "Carr" (marshy land), so it seems unlikely that the bridge was on any long distance east-west packhorse route.

Ravenseat, Swaledale (Group 1) 91: NY 862 034

This single arch alongside a ford across Whitsundale Beck in Upper

The lonely packhorse bridge at Ravenseat in Upper Swaledale

Swaledale is a genuine packhorse bridge in spite of being a few inches wider than 6ft. It lies on the old packhorse road from Barnard Castle to Kirkby Stephen which also served the shallow coal pits around the Tan Hill Inn. These pits were being worked as early as the fourteenth century when owned by Richmond Castle, and later became minor properties in the estates of the formidable Lady Anne Clifford, who, after surviving two husbands, accumulated the castles of Skipton (where she was born), Appleby, Brough, Pendragon and Brougham.

Although a segmental arch, the bridge has a high hump-back and spans 24ft. The width is 79in between 30in parapets which appear to have been added. The roadway surface is of original cobbles. A paved section of the old road can still be seen between the bridge and Tan Hill. Walkers on the Coast to Coast footpath pioneered by A.Wainwright, use this bridge.

Ravensworth (Group 2) 92: NZ 141 081
The minor road to Ravensworth from the A66, crosses Holme Beck

by a rather ugly modern concrete bridge. Immediately alongside on the downstream side is an older bridge (Holme Bridge) now disused, which is sometimes referred to as a packhorse bridge. With a width between parapets of 11ft and with no obvious signs of widening, this description seems improbable. The segmental arch spans about 18ft and the 36in parapets are topped by square copings. The roadway is tarmacked. The bridge is mentioned by Pevsner (*Yorkshire North Riding*).

About a quarter of a mile downstream is another small bridge which has the appearance of being older. It is on a direct line from Ravensworth Castle to the A66 at Fox Well and is similar in construction to the upstream bridge but about 1ft narrower.

An old track for the packhorse transport of lead from Swaledale to Stockton or Yarm for export passes through Ravensworth. Perhaps one or other of these bridges served this route.

Romanby (Group 1) 99: SE 358 935
DOE Listed Bridge. North Yorks County number 66.
Near to Northallerton, this segmental arch spans the Brompton Beck (a feeder of the River Swale) close to a minor road leading from Romanby to Yafforth. The span is 21ft and the width 58in between 18in parapets which are formed from single stone slabs stapled together. The roadway is recent concrete.

This bridge is similar in appearance and construction to the one at Sowerby near Thirsk (q.v.) and may be of the same age. Although there is no obvious packhorse route served by the bridge, there is a Grange Farm just beyond Yafforth, which hints at a Cistercian outpost. The bridge is obviously later than the dissolution of the Monasteries, but the bridge site may not be. Only in recent times have new bridge sites been exploited. Until the twentieth century, it was much more common to repair or replace a bridge at the same crossing point than choose an entirely new site.

Jervoise quotes that the Sessions of October 1620 voted £22 15s 0d to finish repairs, but notes that the present Romanby bridge looks newer than 1620.

Sinnington (Group 3) 100: SE 744 859
This odd little bridge is one of the handful in this guide which now

cross dry land. The shallow segment is of arch-stones only and spans about 10ft. It crosses a dry hollow on the village green. There are no parapets and the overall width is about 6ft 6in. Leading to each end of the bridge is a 20ft long causeway with, as it were, retaining walls built in line with the up and downstream sides of the bridge. The causeway surface is some 6in below the top of these walls.

The bridge was repaired and examined in 1965, this work being reported by John McDonnell in *Ryedale Historian* number 2, April 1966. This paper suggests that the bridge crossed either a mill leat or an overflow channel from the main river, the Seven.

Sowerby (Group 1) 99: SE 435 807
DOE Listed Bridge. North Yorks County number 52.
The bridge crosses the Cod Beck downstream of Sowerby, about 1 mile south of the centre of Thirsk and is alongside a minor road just as it passes underneath the Thirsk by-pass (the A168). The span is 33ft and the width 64in between 29in parapets which appear to have been raised from about 6in. The parapet coping stones, as well as being stapled together across the top, are stapled to the stones underneath.

It is assumed that this bridge was built in 1672 since in that year the Sessions allowed the sum of £10 to be paid as a gratuity towards making Sowerby Bridge "a sufficient horse bridge". It is called Town End Bridge locally and, apart from a ford, provided the only way into Sowerby from the south until 1929.

Stainforth (Group 2) 98: SD 818 672
This is a beautiful segmental arch across the River Ribble. The span is 57ft and the width 6ft 9in between 36in parapets. The bridge, which was built in 1675 and is now owned by the National Trust, carries a road now used by motor traffic and connects Stainforth with the minor road running north-south on the west bank of the river. A track continues west to Feizor.

References to this bridge as a packhorse bridge have more to do with its age than with its proportions, though no doubt plenty of packhorses used it in its early life.

G.N.Wright, in *Roads and Trackways of the Yorkshire Dales* proposes a route joining Ingleton, Clapham, Austwick and Stainforth and then going forward by Mastiles Lane into Upper Wharfedale. Both Clapham (q.v.) and Stainforth bridges could have served this route.

Stokesley (Group 1) 93: NZ 524 085
DOE Listed Bridge. North Yorks County number 196.

The village of Stokesley is about 6 miles south of Middlesborough. The River Leven flows through the middle of the village and within a few yards is crossed by one road bridge, one girder footbridge, two fords and two modern clapper footbridges as well as the packhorse bridge, access to which is down a lane opposite the post office.

The span is 36ft, the arch segmental and the width 64in between 32in stapled parapets. The roadway has the original cobbles.

This seventeenth-century bridge was repaired in 1638 and is on a road from Great Ayton towards the south-west where it probably joined with the famous "Hambledon Drove Road". For a description of this ancient high-level route, see *The Drovers* by K.J.Bonser.

The bridge is known locally as "Taylorson's Bridge", presumably because Taylorson was the builder.

Telfit (Group 2) 92: NZ 088 024
This well-constructed and well-maintained bridge crosses Marske Beck near Telfit Farm and now carries only a public footpath.

The semi-circular arch spans about 24ft and the overall width is 9ft 9in. There are no parapets. Although semi-circular, the arch springs from near stream level and the roadway is at the same elevation as the top of the arch. There is thus no hump-back.

The bridge was built to serve the Clints lead mine nearby on the hillside to the east and traces of the mine road are obvious on both sides of the bridge.

Thornsgill (Group 2) 98: SD 777 794
This odd little bridge spans the River Ribble (which further upstream is called Gayle Beck) about a mile from the Ribble Head viaduct on the Carlisle-Settle Railway. A quarter of a mile north-eastwards on the B6255 is a house called "Gearstones", once an inn, but even earlier the site of an important local market for corn and oatmeal,

active until the latter half of the nineteenth century. About two miles south-east of the bridge at Grid Ref: SD 800 768 is a farm called "High Birkwith" which was also once an Inn, on a packhorse route connecting Settle, Horton-in-Ribblesdale, and Askrigg/Hawes (see also Ling Gill bridge). It is my fancy, without any documentary evidence, that Thornsgill bridge provided access from Gearstones market to the Settle - Horton - Askrigg packhorse road.

Thornsgill is a mini-canyon worn into the limestone which the bridge crosses at a deep and narrow place. The semi-circular arch spans about 15ft and the maximum overall width (no parapets) is 6ft 6in.

The structure was repaired in 1991.

Thornthwaite (Group 1) 104: SE 174 593
DOE Listed Bridge. North Yorks number 214.
This tiny bridge crosses Darley Beck just upstream of a ford on a minor road below Thornthwaite church.

The tiny packhorse bridge at Thornthwaite, North Yorkshire

The bridge has a span of about 14ft and is a mere 38in wide between 24in parapets. Unusually, the parapets are corbelled out from the arch, suggesting an attempt to maintain the original width. The copings are connected together by iron staples. The roadway is paved mostly with flagstones, though there is some modern concrete.

Tucked away in a quiet dell, the bridge has all the imagined qualities of a true packhorse bridge. The arch is segmental, but so nearly a semi-circle that the result is a satisfyingly high hump-back.

The bridge is said to be on a track joining Ilkley and Ripon.

Uckerby (Group 3) 93: NZ 251 015

This crudely-built segmental arch of 10ft span and 10ft overall width (no parapets), springs from about 3ft above the normal level of Scorton Beck. Only the sandstone arch-stones remain, any ramps leading to the arch having long since gone.

There is a Holy Well - "St Cuthbert's Well" - in the adjacent field. Just downstream, an interesting horseshoe-shaped bridge once carried a railway line (now abandoned) across the beck.

The most likely purpose of Uckerby Bridge, was to carry farm traffic. It presently seems to be restricted to use by stock crossing the beck.

Wath (Group 2) 99: SE 144 677

This village in Nidderdale has a name which is itself the Old Norse word for a ford. The bridge crosses the River Nidd just outside the village in one arch plus an overflow arch. The span is about 51ft and the present width 8ft 6in. Until 1890, the bridge was 5ft wide and thus a true packhorse bridge. The widening, on the upstream side, is obvious underneath the bridge.

The name confirms that this was a crossing point of the River Nidd before a bridge was built. Perhaps the ford and later, the packhorse bridge, served to connect farms on the west bank of the River Nidd with markets at Ripon and Kirkby Malzeard. An additional possibility is that it was built before burial was permitted at Middlesmoor, so that corpses could be carried to the Parish Church also at Kirkby Malzeard.

West Burton (Group 2) 98: SE 019 868

Burton Bridge crosses Walden Beck just downstream of the village proper and connects the green lane known as Morpeth Gate to the B6160. At 10ft wide between 3ft parapets and with no sign of widening, the bridge is wide enough for carts and presently carries farm traffic. However, most authorities agree its status as a packhorse bridge and since there was a lead smelting mill at West Burton, its main function was no doubt to serve the packhorse trains carrying lead ore to the mill.

From the south, a 3ft wide causeway rises in a gentle ramp parallel to the road and joins the west end of the bridge. From the road at right angles to the ramp are pedestrian steps. The bridge spans about 34ft, has a segmental arch and a "YNR" plate on the downstream parapet.

Upstream and just below West Burton waterfall is a single arch spanning 24ft. This bridge is 42in wide between 32in parapets and thus looks more like a packhorse bridge than does Burton Bridge. Adding to the uncertainty, on the east bank of the river, away from the village, a paved causeway ascend the steep river bank. This bridge however, is a footbridge built in 1860.

Westerdale (Group 2) 94: NZ 662 062

DOE Listed Bridge. North Yorks County number 192.

This bridge is known as Hunter's Sty Bridge (Sty or Stee from an Old Norse word meaning a path) and is the best example I know for seeing ribbed construction from above. Across the ribs of the very high hump, slabs are laid acting as beams and the river can be seen through the cracks between the slabs.

The span is 18ft and the width 9ft 8in between 43in parapets. On the outside of the upstream parapet a coat of arms is carved and on the outside of the downstream parapet a quotation is incised which reads "THIS ANCIENT BRIDGE WAS RESTORED BY COLONEL THE HONOURABLE G DUNCOMB A D 1874".

The bridge is thought to have been built before the end of the thirteenth century and, according to Pevsner, the restoration of 1874 was confined to the parapets, though Jervoise writes that the bridge was "too thoroughly restored in 1874".

Hunter's Sty Bridge near Westerdale, North Yorkshire

Yockenthwaite (Group 2) 98: SD 905 790
Another of the lovely single-arch bridges in North Yorkshire (see Stainforth, and Ivelet) which are too wide to be restricted to horses only but are generally called packhorse bridges. Yockenthwaite crosses the River Wharfe in a single span of about 54ft. The width is 7ft 6in and the bridge presently provides vehicle access to the hamlet.

The B6160 from West Burton along Bishopdale, connects with Wharfedale. However, at Cray, a mile before the descent to Buckden, a track continues at high level above Hubberholme before descending to Yockenthwaite. Across the bridge, this track then continues over Horse Head Moor to Halton Gill where it splits. One branch heads for Horton in Ribblesdale, by-passing Pen-y-Ghent on the north side; the other aims for Stainforth and Settle, passing Pen-y-Ghent on the south. The bridge would be useful for pack trains travelling between Bishopdale and either Horton or Settle.

SOUTH YORKSHIRE

Longshaw (Group 1) 110: SK 263 815

Sometimes called the Burbage Brook Bridge, this small arch spans Burbage Brook in the shadow of what looks like a Dartmoor Tor. This dramatic outcrop of shattered gritstone blocks is Carl Wark with a hill fort on its summit. Although undated, current opinion seems to favour a defensive work built by native Britons against invading Anglians from Mercia, perhaps as late as the seventh century.

Although the surrounding hillsides are scattered about with boulders of millstone grit, the bridge is located where the stream has worn into a band of dark coloured, soft and flaky shale. It may be that this unstable material has allowed the abutments to move slightly, for although the arch is segmental when viewed from upstream, it has twisted and appears as a shallow point from downstream. The twist has allowed one arch-stone near the top on the downstream side to drop almost to the point of falling out.

The arch is of unusual construction. Viewed from the side, each arch-stone is about 12in by 6in, but from underneath, each course is of two stones only; a long (about 4ft) and a short (about 1ft), laid alternately.

The bridge spans about 12ft and is 60in wide overall (no parapets). The roadway is paved with gritstone slabs averaging about 36in long, by 18in wide by 6in thick, laid across the roadway - a very robust wearing surface (see also Beresford Dale).

Two ancient "saltways" from Cheshire via Macclesfield approach Sheffield from the west; the northernmost via Chapel-en-le-Frith, the Hope Valley, Stanage Edge and "The Long Causeway", and the southern via Buxton, Tideswell, the Fox House Inn and Ringinglow. Both these routes were used into the nineteenth century by packhorse trains carrying Sheffield cutlery in the opposite direction. The footpath which uses Longshaw Bridge leaves this southern route at Grid Ref: SK 276 817 and joins the northern route at Hathersage. It thus serves to interconnect important alternative routes leaving Sheffield for the west. (See also the Rivelin Valley Bridge.)

Also, the bridge is almost due west and only four miles from Abbeydale which, although now an industrial museum, was an active scythe smithy in the early years of the eighteenth century.

Although well within the Peak District National Park boundary, the bridge is about half a mile inside South Yorkshire.

Oxspring (Group 1) 110: SE 268 027

Called locally "Willow Bridge", this beautifully proportioned high arch crosses the River Don less than 2 miles downstream from Penistone. The span is 30ft and the width between parapets 39in. The parapets are 38in high with stapled copings and the roadway cobbles have at some time been infilled.

From the south, a walled and sunken lane, the remnant of an old packhorse road from Bradford, leads to the bridge from the B6462. Across the river the route once headed north-north-west for Wakefield by way of Silkstone. See also "Unsliven" and "Wharncliffe Side".

Rivelin Valley (Group 1) 110: SK 293 873

This arch, which has a shape suggestive of a semi-ellipse, crosses the River Rivelin at the end of the "Coppice Road". It is about 4 miles west of Sheffield just off the A6101 near its junction with the A57. The span is 22ft and the width 37in between 23in parapets. The parapet coping stones are joined by leaded-in iron staples and the whole construction is of dressed sandstone.

Many packhorse routes radiate from Sheffield towards the west, amongst them, one on either side of the Rivelin valley. Just to the north there is a route over Rod Moor towards the Woodlands valley, to the south "The Long Causeway" towards Hathersage. The Rivelin valley bridge is against the grain of these routes, almost as though it were part of the packhorse equivalent of a modern ring road.

Unsliven (Group 2) 110: SK 254 992

This is now a substantial minor road bridge more than 20ft wide, but evidence of its packhorse origins can still be seen. The span is about 25ft across the Little Don River immediately below the dam and outlet works of Underbank reservoir.

D.Hey, in *Packmen, Carriers and Packhorse Roads*, writes that the bridge, on the packhorse road from Bradfield to Wakefield, was wooden in about 1730 and that the first stone bridge was built shortly afterwards (see also Oxspring and Wharncliffe Side which are or were on the same route).

Three distinct arches can be seen underneath and it appears that the central section (about one quarter of the total width) was the first. The upstream section must be the first widening (again about one quarter of the total width), because on a coping stone at the north end of the upstream parapet is carved "July 1796". The downstream half of the bridge is more modern and massively constructed in gritstone. The parapets are about 4ft high, topped by triangular copings.

Wharncliffe Side (Group 2) 110: SK 292 943

This is another rebuilt bridge which was to have been drowned under a reservoir. Originally, it was called "New Mill Bridge" and crossed Ewden Beck under what is now Ewden reservoir. Built originally in 1734 to serve a route joining Sheffield and Leeds via Bradfield, Ewden, Silkstone and Wakefield, it was saved from destruction and rebuilt on its present site by a local paper

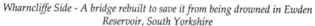

Wharncliffe Side - A bridge rebuilt to save it from being drowned in Ewden Reservoir, South Yorkshire

manufacturer. It now crosses the Wharncliffe Beck as part of a scenic walk through Glen Howe Park. The span is 22ft and the width 56in between 15in parapets.

The route served by this bridge before its removal and rebuilding also made use of the bridges at "Unsliven" and "Oxspring" (q.v.).

Whiston (Group 3) 120: SK 440 898
Located in a small and apparently forgotten rectangle of land bordered by the M1, the A631, the A618 and the A630, this small bridge crosses a tiny brook just before it joins the River Rother. The span is 15ft and the width 10ft 6in. There are no parapets and the bridge is formed of arch-stones only.

About a mile south-east of this bridge, the A618 crosses a stream that is now the outflow from Ulley reservoir, a place apparently still known as "Packman Bridge". This argues for a packhorse route to the south from Rotherham, of which Whiston Bridge may have been part. On the other hand, the bridge, interesting and old-looking though it is, may be simply for access to farm land.

WEST YORKSHIRE
Alcomden (Group 2) 103: SD 956 321
Alcomden Bridge, at a spot named Holme Ends, crosses Alcomden Water about half a mile downstream of the dam of Walshaw Dean Lower Reservoir. Holme Ends now consists only of a ruined barn, but the site and the bridge are on the line of "Limers Gate", an old packhorse route connecting sources of lime to the north-west, with the Halifax area. The old packhorse road is best seen as it contours round New Laithe Moor, south and east from the bridge. In other directions, access roads for the building of the three Walshaw Dean reservoirs now dominate and masonry channelling the stream upstream of the bridge, has the appearance of "waterworks" construction.

The bridge has two substantial semi-circular arches which span about 30ft. The width is 10ft between 32in-high parapets which have copings of large gritstone slabs about 4ft long, by 16in wide, by 6in thick, laid flat. The midstream pier has low level triangular cutwaters both upstream and downstream and the eastern arch is dry at normal stream flows.

Lumb Bridge, Crimsworth Dean (q.v.) is also on "Limers Gate".

Beaumont Clough (Group 2) 103: SD 980 261

This simple, parapetless arch across the stream in Beaumont Clough, serves an old packhorse road joining Hebden Bridge and Rochdale. Many clues remain. About half way between the bridge and Hebden Bridge, is a farm called "Horsehold". The route contours round the bottom of the hillside under Stoodley Pike (where it is called "London Road") to Mankinholes where there is a well-preserved horse trough, there to join the packhorse track from Crag Vale. Having crossed the valley at Todmorden, a possible line continues via Ragby Bridge (q.v.): the Long Causeway (one of many Long Causeways) and the Ramsden Road which heads straight into Watergrove Reservoir. The route from the reservoir to Rochdale is mostly lost under building development.

The bridge spans only 4 or 5ft and has a width of 96in. The roadway is now grass-grown and the bridge constructed without mortar. Walkers travelling north on the Pennine Way who choose to leave the official route to visit Hebden Bridge might do so by crossing this beguiling bridge.

Bingley (Beckfoot Bridge) (Group 1) 104: SE 103 385

This splendid old bridge was built in 1723. Two masons, Benjamin Craven and Josiah Scott were contracted to both build the bridge and keep it in good repair for seven years. For this they were paid £10. They built well, the bridge is still in good shape, with a little obviously newer pointing and work to the footings of the abutments.

The bridge crosses Harden Beck in a single segmental arch, on the line of Beckfoot Lane, an old track which follows the south bank of the River Aire and was once a main north-south route. The present Bingley main road bridges (Ireland and Cottingley) were both built later than Beckfoot Bridge.

The span is about 33ft and the roadway which is mostly cobbled but with a few flags, is 62in wide overall (there are no parapets). Some arch-stones appear to be the full width of the bridge, other courses are made up of long and short stones.

Leaning against the southern upstream abutment is an old

millstone, above which are carved what look like Mason's Marks. These are disposed vertically and from the top read - "W" (with a horizontal mark above): "L" and "V" (with a horizontal mark beneath. The bridge is disfigured by heavy wooden railings which have the same appearance as those on a photograph published in 1898.

Just upstream, there is an interesting footbridge, probably built for access to the mill which was once on the site.

Clayton West (Group 1) 110: SE 261 117
This semi-circular arch crosses the River Dearne just outside the village. It leads from the A636 into the yard of a mill and is known as Park Mill Bridge. The bridge has a span of 24ft and is 55in wide between parapets. The original kerbs were 11in high but have been increased to 25in using a different stone. The approach from the A636 shows signs of cobbles under the grass but the bridge roadway is paved with stone slabs (flags) which rise to a shallow point at the centre.

From the bridge, a line of lanes and minor roads leading north-west towards Huddersfield via Emley and Grange Moor is evident. Across the bridge towards the south-east footpaths and minor roads via Hoyland church and Cawthorne lead to Silkstone. Perhaps Park Mill Bridge served a Huddersfield variation on the Bradfield - Wakefield packhorse route mentioned under Oxspring (q.v.).

Crimsworth Dean (Lumb Bridge) (Group 1) 103: SD 992 314
The word "lumb" (see also Lumbfoot Bridge) has an interesting derivation. Originally meaning "a well for collection of water in a mine", it seems to have evolved to mean a pool. Lumb Bridge is just upstream of Lumb Hole waterfall which drops into a lovely tree-shaded pool. It is altogether an idyllic spot. Walled packhorse tracks lead steeply down the western and eastern hillsides to Crimsworth Dean Beck and the bridge. Worth noting is the Norse word "beck", unusual in these parts.

The packhorse routes using Lumb Bridge probably included tracks between Hebden Bridge and Keighley, Hebden Bridge and Burnley via Gorple Gate, and "Limers Gate" leading towards Halifax from the north-west. This last possibly made use of four bridges described in this guide - Higherford, Catlow Bottoms, Alcomden, and Lumb.

The bridge itself is segmental but almost a semi-circle and built from gritty sandstone decorated by patches of a brick-red lichen. There are no parapets and the overall width is 60in, the span 15ft. The centre of the span comprises arch-stones only, the roadway has been concreted and the bridge is disfigured by iron railings.

Eastwood (Group 2) 103: SD 964 253

The Calder valley between Todmorden and Hebden Bridge is narrow and fully occupied by a main road, a railway, a canal and the River Calder. What little flat land remains is mostly taken up by old industrial buildings. This being Yorkshire however, space has been set aside for a cricket field (Eastwood). Crossing the river, near the cricket pavilion, is a bridge which could be used in a textbook to illustrate the construction of a shallow, segmental arch.

Built of arch-stones only, the span is about 20ft and the overall width (no parapets) is 13ft.

It is difficult to propose a route for this bridge, which indeed is wide enough to take carts. It looks old and, if built before the canal, may have provided a short-cut from Upper Eastwood on the north side of the valley, to ascend the little clough opposite and join the "London Road" described under Beaumont Clough (q.v.).

According to the groundsman at the cricket field, the bridge is now mainly used by people walking the Calderdale Way in day-long bits, crossing to use the toilet or buy a cup of tea at the pavilion on Saturday afternoons.

Hawks Clough, Mytholmroyd (Group 3) 104: SE 007 263

Interesting though it is, Hawks Clough bridge fails to meet the criteria for a packhorse bridge, except perhaps age. It spans the River Calder about 500yds west of Mytholmroyd Bridge, immediately alongside the A646. The arch is segmental,massively built of gritstone and spans about 60ft. The roadway is 10ft wide between substantially built parapets 34in high.

Although sometimes referred to as a packhorse bridge, the southern end of the bridge, away from the main road, is defended by massive gateposts which still carry ironwork fixings for hanging a pair of gates. Wide enough for carts, it has all the appearance of a mill access bridge, although no mill remains. It is now used only by pedestrians.

Haworth (Long Bridge) (Group 1) 104: SE 020 376

Long Bridge presents an odd, isolated appearance. The arch springs from stream level and although a segment, it is nearly a semi-circle. Instead of a ramp leading to the arch, there are three steps at each end and these surmount very small abutments. The parapets, which are 25in high in the middle, increase to 45in high at the ends, where they are finished off by "gateposts"; thus the parapet tops in silhouette appear horizontal. All these constructional details give an impression that the bridge has been "plonked down" whole. It is imposed on its location rather than growing from it.

The bridge crosses the River Worth in a span of about 27ft and is 32in wide between parapets which are topped by blocks stapled together. There is a modern concrete ford upstream.

Long Bridge is so near to Lumbfoot Bridge (q.v.) that it is difficult to understand why two bridges were built. They both connect the south-facing slopes of the Worth valley with Haworth, though Long Bridge has the more direct route via named lanes (Hey Lane and Street Lane). Lumbfoot has traces of an industrial past, so perhaps the hamlet needed its own bridge.

There is a possible longer route via Pickles Hill, White Lane and Goose Eye in the general direction of Keighley, which both bridges may have served, though for Long Bridge to have carried packhorses, the existing steps must have been covered by a ramped causeway.

Hebden Bridge (Group 2) 103: SD 993 273

DOE Listed Bridge. West Yorkshire County number 53.
This ("Hepton Brig") is the bridge from which the town presumably gets its name. It was built in 1510 to replace one of wood. Repairs were carried out in 1602 and 1657 and the parapets were repaired in 1845 and raised in 1890. There are three arches (one dry) crossing Hebden Water in the middle of the town, with a total span of about 60ft. The parapets are presently 45in high and the roadway 96in wide between them. Both piers have upstream and downstream cutwaters three of which are carried up to form pedestrian refuges

now railed off. The roadway is cobbled, the arch has quite a steep hump, and the bridge is now used only by pedestrians.

Hebden Bridge is at the heart of that ribbon of early industrial development along the valley of the River Calder and was a junction of many packhorse routes connecting Lancashire and Yorkshire.

Hebden Bridge (Foster Lane) (Group 2) 103: SD 992 278

An alternative name for this bridge is "Hollins Lane" and local opinion (ie. a lady living nearby) is (a) upset at insensitive development (a new wall and a lamp standard) spoiling the appearance of the bridge, (b) feels that the bridge needs the protection of a DOE listing, and (c) thinks it might be sixteenth century - see below. Heartening to find concerned local people.

Many Yorkshire packhorse bridges share the general appearance of this bridge - a humped arch, but with parapets rising to a shallow point. The span across Hebden Water is about 33ft and the width 75in between 27in parapets. The parapet copings are joined by leaded-in iron staples and the roadway at the centre of the arch is now protected by modern cobbles set in concrete.

The bridge serves to connect Hebden Bridge and Heptonstall on the hill top to the west. Heptonstall has the oldest remaining Cloth Hall in the district (1545), so the bridge might indeed be sixteenth century.

Hudson's Bridge (Group 2) 103: SD 922 266

Hudson's Bridge is on one of the many possible continuations south-eastwards (to Lydgate?) of the Long Causeway, a medieval route joining Burnley and Halifax.

The bridge spans 6ft across Redmires water and is 78in wide between 24in parapets. Built using dressed gritstone, the parapet stones are large slabs and at the western end, inboard of a modern gate, are two slotted "gateposts" (for dropping in a hurdle?). From the west, the approach is a holloway, paved in part by stone flags.

Of the many possible derivations of "Redmire" as a name for the stream crossed by this bridge, I favour "mere or lake where reeds grow" - note the area around the disused Redmires dam upstream.

Lower Strines - One of the cluster of small packhorse bridges around Hebden Bridge, West Yorkshire

Lower Strines (Group 1) 103: SD 959 285

This little bridge near Lower Strines farm, crosses Colden Water a few hundred yards upstream of Jack Bridge. It is one of several West Yorkshire packhorse bridges which would nowadays be difficult for a packhorse to negotiate. Lower Strines Bridge has a restricting stile at each end, Oxenhope and Haworth - Long Bridge (q.v.) each have steps; others have high parapets.

The arch is segmental with a high hump and spans about 22ft. The width is 32in between parapets built with six gritstone slabs which are 18in high and were once stapled together; sockets full of lead remain, the iron staples have gone. The roadway is made up of flags except for the centre 6ft or so which has a thin skin of concrete over the arch-stones.

A narrow, grass-grown, cobbled causeway leads down to the bridge from east and the probability is that bridge served to connect the south-facing slopes of the middle reaches of the Colden valley to Blackshaw Head. From here, "The Long Causeway" heads west for Burnley, and "Badger Lane" east to Hebden Bridge, both part of an important east-west packhorse road.

Lumb Foot (Group 1) 104: SE 015 375

This little bridge, now made superfluous by a wide farm access bridge, crosses the River Worth a couple of miles west of Haworth. Although carrying a footpath (not a bridleway), it connects Haworth and Stanbury to the complex of lanes and hamlets which fill the south-facing slopes of this upper length of the Worth Valley. Also, there are signs of one-time small-scale industry in Lumbfoot itself.

The bridge is a shallow segment springing from 5 or 6ft above normal stream level and spans about 23ft. The width is 31in between parapets made from single slabs 19in high which rise to a shallow point in the centre and are stapled together.

There is little chance of this bridge being overlooked because the parapets are topped by iron railings painted a very bright green.

See also Haworth - Long Bridge, less than half a mile downstream.

Marsden (Closegate) (Group 1) 110: SE 029 121

DOE Listed Bridge. West Yorkshire County number 69.
The feeder streams of the River Colne rise on Close Moss about two miles west of Marsden and Closegate Bridge crosses the stream which is now the outflow from March Haigh Reservoir.

The arch is segmental, spans about 18ft and the width between 24in parapets is 50in. The parapets are topped by rounded copings which were once stapled together, though the staples have gone, leaving the lead-filled sockets.

The hamlet of Hey Green, just downstream, had, in 1710, a water-powered Fulling mill and a field on the south of the river still contains rows of stone tenter posts. Closegate Bridge doubtless supported more packhorse loads of woollen cloth than any other commodity.

The Closegate packhorse track from the bridge can still be followed and crosses the A640 at its highest point, where it coincides with the Pennine Way long distance footpath. Between the bridge and the A640 and near to the main road, there is a stone post lettered "P H ROAD". The route, known also as "Easter Gate" or "Th'owd Gate", linked Marsden with Rochdale.

Hokenhull Centre and East bridges. There are three bridges in all at this Cheshire site - a must for enthusiasts.
West Rasen Bridge, Lincolnshire

Chew Stoke Bridge, Avon *(A.Menarry)*
Chewton Keynsham Bridge, Avon *(A.Menarry)*

Marsden (Closegate) - On the packhorse track known as "Easter Gate",
West Yorkshire

Marsden (Village) (Group 1) 110: SE 046 117

This well-maintained bridge with a segmental arch crosses the River Colne just behind Marsden church. In *The Bridges of Lancashire and Yorkshire*, Margaret Slack speculates that it might have been used by packhorses carrying corn to and from a nearby Manorial corn mill. Additionally, it would have been useful to link with the old packhorse route between Marsden and Littleborough/Rochdale which also crosses Closegate Bridge (q.v.) and is known as "Easter Gate" or "Th'owd Gate".

The Marsden village bridge has a span of about 33ft and the width between 34in parapets is 36in. Rounded coping stones top the parapet and the roadway is cobbled.

Oxenhope (North Ives) (Group 1) 104: SE 036 360

This little bridge lies on a bridleway between Oxenhope and Haworth which passes under the Worth Valley railway immediately after crossing the bridge.

Pevsner *(West Yorkshire)* describes the bridge under "Oxenhope" as a packhorse bridge, though the three steps at each end raise the question of laden packhorses climbing steps; perhaps they are a later addition or original covering ramps have gone.

The segmental arch spans 20ft and the width is only 25in between single-slab, 16in high parapets, a few of which are stapled together and all of which sport a healthy growth of mosses and lichens. As on many of the small bridges in West Yorkshire, the parapets rise to a shallow point in the centre.

The roadway is made up with well-laid and only slightly worn sandstone flags. These appear much newer than the bridge and I speculate that the steps were perhaps built when the flags were laid.

The bridge crosses Bridgehouse Beck and is known locally as "Donkey Bridge".

Oxygrains (Group 2) 110: SE 004 158

DOE Listed Bridge. West Yorkshire County number 142.

This old bridge is within both sight and earshot of the frantic traffic on the M62 and the contrast could not be more stark. It crosses the outlet stream from Green Withens reservoir below the massive masonry road bridge which carries the A672 across the same stream 100 yards

Oxygrains - Within earshot of the M62, West Yorkshire

away but at a higher level.

The bridge, of segmental shape, is composed of massive gritstone arch-stones only and spans about 16ft. Although 8ft 6in wide overall (there are no kerbs or parapets) it is officially designated a packhorse bridge by the DOE. Certainly, it would be a brave man who tried to drive a coach or a cart across in its present state, and if once there were parapets, the width would be reduced to packhorse bridge proportions.

It is possible that the route now followed by the A672 over Windy Hill, was once an old road joining Failsworth with the Calder valley via Ripponden.

Ragby Bridge (Group 2) 103: SD 922 216

Ragby Bridge crosses the stream which flows down Ramsden Clough from Ramsden Clough Reservoir. The clough is steep, with waterfalls both upstream and downstream of the bridge. Walled lanes lead to the bridge from Todmorden in the north. Southwards, the lane is named "The Long Causeway" and I speculate that it is on a route between Hebden Bridge and Rochdale more fully described under "Beaumont Clough" (q.v.). From the north the track is a distinct holloway.

The bridge is presently used by farm traffic and has been well-maintained recently, including the addition of low concrete parapet walls.

The bridge spans 6ft and is 9ft wide between the parapets.

Ripponden (Group 2) 110: SE 041 198
DOE Listed Bridge. West Yorkshire County number 215.
This single arch across the River Ryburn is alongside the modern road bridge on the Elland road, near the centre of the town. The span is about 60ft and the width between 30in parapets is 9ft 6in. The road surface is cobbled.

The bridge is used by cars to gain access to adjacent property and has a Road Traffic Acts weight limit of 5 tons. The hump is steep and the cobbles at the centre of the span bear many scars from the undersides of passing vehicles.

The bridge was built in 1533 and restored in 1973, though the adjacent Bridge Inn (originally a religious establishment) probably dates from the fourteenth century, suggesting an even earlier crossing place.

Standedge (Thieves Bridge) (Group 2) 110: SE 021 101
This massively constructed, parapetless arch is in as wild a spot as any bridge I know (packhorse or not). High on the moors of Standedge Edge, to the north-west of the summit cutting which carries the A62, it crosses Thieves Clough so high up that it has hardly yet become a clough.

The bridge has a span of about 12ft and is 9ft wide overall. It is built from large blocks of blackened gritstone, and unusually, similar masonry is used to direct a bend in the stream through the arch. Margaret Slack, in *The Bridges of Lancashire and Yorkshire*, considers it was built about 1760 and is unlikely to be a packhorse bridge despite its appearance.

On the other hand, W.B.Crump in his paper "Saltways from the Cheshire Wiches" describes a section of route (his SALTWAY E) crossing a Salter Hebble in Delph before traversing Standedge by Thieves Bridge on its way down to Marsden via the south side of Pule Hill.

The turnpike from Austerlands (Oldham) to Huddersfield dates from 1759, before the cutting which carries the present A62, though it seems unlikely that a turnpike trust would build a bridge restricted to 9ft in width.

Thieves Bridge is just one of the many instances where questions remain concerning the status of old bridges.

Wakefield (Group 1) 104: SE 338 201

This oddly located bridge, near to the centre of Wakefield is now totally on dry land, many feet above the present level of the River Calder. It is sandwiched between the southern end of the well-known fourteenth-century Chantry Chapel bridge and a garage.

The three arches which approximate to semi-ellipses, span 51ft, the width between parapets is 61in, and the parapets are 30in high with flat stone copings. The bridge was beautifully built of dressed stone in 1730 and is well maintained. In a sense, it provides a continuation of the fourteenth-century Chapel Bridge, presumably useful when the level of the River Calder was high.

Unexpected packhorse bridge near the middle of Wakefield, West Yorkshire

THE BRIDGES OF THE MIDLANDS REGION

BEDFORDSHIRE
Sutton

CHESHIRE
Hokenhull
Marple

DERBYSHIRE
Alport
Ashford-in-the-Water
Bakewell
Barber Booth
Beresford Dale
Edale
Goyts Bridge
Haddon Hall
Hayfield
Hollinsclough
Milldale
Raper Lodge
Slipperystones
Three Shires Head
Washgate Bridge
Youlgreave

GLOUCESTERSHIRE
Bibury
Lower Slaughter
Slad

HEREFORD & WORCESTERSHIRE
Shell

LEICESTERSHIRE
Anstey (Village)
Anstey
Aylestone
Bottesford
Ederby
Medbourne
Rearsby
Thurcaston

LINCOLNSHIRE
Scredington
Utterby
West Rasen

NORFOLK
Houghton St Giles
Walsingham

NORTHAMPTONSHIRE
Charwelton

SHROPSHIRE
Rushbury

STAFFORDSHIRE
Great Haywood

SUFFOLK
Cavenham
Moulton

WARWICKSHIRE
Tidmington

WEST MIDLANDS
Hampton-in-Arden

INTRODUCTION TO THE MIDLANDS REGION

Nearly one third of the packhorse bridges in the Midlands Region are concentrated in the Peak District of Derbyshire, most of which is designated a National Park. Ringed by industrial cities, this most visited of the National Parks suffers millions of tourists every year, but there are still remote places and many traces of packhorse activity.

Holloways, some of them walled, are still to be found, especially on steep hillsides leading to river crossings. A scrutiny of the Ordnance Survey 1 inch tourist map of the district reveals more than twenty "crosses" and though many of these are now simply place names and others refer to village preaching or market crosses, some at least are thought to be medieval, marking boundaries and/or trackways; Edale Cross at Grid Reference SK 077 861 is a good example. In the late seventeenth century, an Act of Parliament required guide posts to be erected at crossroads between market towns. Many of these are still in place and a dozen or so of them on the high land between the River Derwent and the industrial belt to the east are listed in D.Hey's *Packmen, Carriers, and Packhorse Roads.*

Of the dated bridges, the earliest is 1664 and the latest 1734. This corresponds well with the period 1660 to 1760 proposed by Collingwood for the bridges in the Lake District.

Packhorse transport was at its busiest when the towns and cities which surrounded the park were beginning their industrial expansion and the two events were of course interdependent. There was thus much through traffic and a few of the tracks can still be followed significant distances on foot. Many of these routes are described in detail in *Peakland Roads and Trackways* by A.E. and E.M.Dodd.

Of significant cargoes in the Peak District, lead was important (it has been mined in Derbyshire since Roman times); chert was carried from quarries near Bakewell to the potteries for use in glazes; and tobacco was carried from Liverpool across the district, to industrial areas to the east.

Although this guide does not include clam or clapper bridges, it is worth noting that Dartmoor is not the only area where they exist

Edale Cross - A medieval route marker at Grid Ref SK 077 861 in Derbyshire

and were used by packhorses. On Big Moor, west-north-west of Chesterfield, two clam bridges span the Bar Brook. One, at Grid Reference SK 276 782 has an attendant guide post and is near the medieval Lady's Cross, thought to be both a boundary and track marker. The other, at Grid Reference SK 278 741, is dated 1742. Both are packhorse crossings rather than footbridges.

Unlike Derbyshire, the other counties which make up the Midlands Region are without significant high ground and the gentler terrain encouraged the replacement of packhorse transport by carts and wagons at an early date. In turn this led to the replacement or widening of many narrow bridges.

Through the years, increasing urbanisation, intensive farming and improved drainage have reduced and channelled some of the sluggish streams of the Midlands. The bridges at Charwelton and West Rasen, for example, now span trickles which a toddler could paddle across; it seems a paradox that there should be bridges there at all. But our forebears were not so profligate as to build unnecessary bridges, so these streams must once have been obstacles to packhorse traffic.

In the Middle Ages, the Shire counties of Nottingham, Leicester, Lincoln and Northampton were famous for raising "Longwool" sheep to satisfy the hunger of foreign merchants for fleeces with a long staple. There was thus both the motivation and the prosperity to build well-found bridges to help the packhorse-borne wool on its way to the ports. This probably helps explain the fact that nine of the bridges described are fifteenth century or earlier. It is fortunate that the subsequent increase in wheeled traffic did not lead to the total destruction or modification of all these old bridges.

The goods most likely to be on board a packhorse train in the Midlands generally, would be corn, from east to west, wool and cloth to the ports and important markets and the ubiquitous salt from the evaporating pans at Droitwich and in Cheshire.

*The massively-built packhorse bridge at Sutton in Bedfordshire.
Originally founded on elm timbers*

BEDFORDSHIRE

Sutton (Group 1) 153: TL 221 475
DOE Listed Bridge. Bedford County number 9.
This massively constructed bridge probably dates from the thirteenth or fourteenth century. There are two unequal pointed arches with a total span of 27ft. The width is 72in between 29in parapets which have triangular copings and the single upstream cutwater has a pedestrian refuge. The bridge is built from dressed sandstone and there is a string course at the base of the parapets which suggest that they may have been added at a later date. The bridge spans Potton Brook which flows into the River Ivel and there is a ford alongside on the downstream side.

Sutton is only a couple of miles east of Ermine Street (The Great North Road) and the bridge may have provided access to this from the east. No other long-distance packhorse route is obvious.

In conjunction with repairs in 1986, the bridge construction was surveyed by a professional archaeologist and the results published in *Bedfordshire Archaeology*, Vol 18 - "Sutton Packhorse Bridge", by Peter McKeague. Of interest from this work in the context of this guide, is the fact that the original bridge was founded on elm timbers laid across the stream. The only other old bridge which was founded on timber and known to me, is "Whitemill Bridge" at Sturminster Marshall in Dorset. This was founded on oak rafts laid on oak piles.

CHESHIRE

Hokenhull (Group 1) 117: SJ 477 657

For anyone interested in packhorse bridges, this site is a must. The approach from the east is down a remote lane which narrows to a footpath, with a distinct holloway alongside for several hundred yards. A bend in the path reveals not one, but three bridges, linked by paved and cobbled causeways.

Jervoise writes that "Among the gifts from the Black Prince in September 1353, was the sum of 20s for the repair of the bridge of Hokenhull, which may have been the site of the present packhorse bridges." He considers that these date from the end of the seventeenth century.

The easternmost bridge, which does not now span a permanent watercourse, is segmental in shape with an 18ft span and is 59in wide between 28in parapets topped by chamfered squ．re copings which are stapled together. The roadway is cobbled. Each buttress is guarded by a triangular cutwater, the purpose of which is puzzling since they are on the downstream side.

From this bridge, a causeway about 35yds long leads to the middle bridge, which spans the present course of the River Gowy. This almost semi-circular segmental arch spans about 21ft and is also 59in wide between parapets. There is much new concrete in the stream bed both underneath and alongside this bridge.

A causeway about 65yds long leads to the third bridge which also spans about 21ft and is 59in between parapets. Again there is now no permanent watercourse under this westernmost bridge. At each end of the arch the downstream parapet is stepped out to a width of about 86in.

All three bridges are beautifully built of well-dressed red sandstone to the same design.

The bridges are probably on a salt route from Nantwich to Chester (there is a Salterwell House just west of Tarporley - en route). Local use is suggested by a "Walk Mill", shown less than a mile upstream on older maps. ("Walk mill" is the old name for a fulling mill.)

Marple (Group 1) 109: SJ 967 873

This graceful arch across the River Goyt is one of the many called "Roman", though the best estimation of a building date is mid eighteenth century. The designation "Roman" also applies to the lakes a little further downstream, no doubt part of the same romantic invention. In its early life it went by the name "Windy Bottom Bridge" and was probably used by farmers en route to Stockport from the high ground around Mellor and Mellor Moor.

The bridge has a span of 45ft and an overall width of 69in. Although segmental, the arch springs from near river level, so the hump is high and at the centre is composed of arch-stones only. Although there were once parapets, these no longer exist and have been replaced by iron railings.

DERBYSHIRE

Alport (Group 3) 119: SK 221 645

Alport is the site of a ford of great antiquity, used much later by packhorses on the very busy Derby to Stockport route. By the end of the seventeenth century "great gangs of London carriers and drifts of malt horses" were using the crossing and in times of flood were being badly delayed. In 1718 the Sessions ordered a "Horse Bridge" to be built but this was replaced by the present bridge, built in 1793, which has a roadway 18ft wide.

A short distance upstream of this bridge is a narrow bridge described in a paper by the Derbyshire County Surveyor as a packhorse bridge. However, in *Peakland Roads and Trackways* by A.E. and E.M.Dodd, the bridge is described as a footbridge, built for private access. There is now a very narrow stile at each end of the

One of the many packhorse bridges known as "Roman Bridge".
At Marple in Cheshire

bridge which would certainly exclude packhorses. It is though, an interesting bridge with a single arch plus an overflow arch with a total span of 21ft. The width is 34in between 30in parapets.

It is the importance of Alport in the history of packhorse transport, plus the interesting small bridge which prompts inclusion amongst these Group 3 bridges.

Ashford-in-the-Water (Group 2) 119: SK 195 697
DOE Listed Bridge. Derbyshire County number 52.
As the name suggests, Ashford is the site of an ancient ford across the River Wye, its importance as a crossing point confirmed by the three present bridges. Ashford Bridge was reported in 1776 as being "so extremely narrow that it is dangerous for carriages to pass over the same". This often quoted report is sometimes applied to the "Sheepwash" bridge, but more probably refers to the bridge on the road to Bakewell which is dated 1664.

It is the Sheepwash Bridge which attracts DOE listing and which is often referred to as a packhorse bridge. It has been widened at some time to the present width of 15ft 4in between parapets. Sheepwash Bridge is nonetheless a lovely structure, with three very shallow segmental arches spanning a total of about 60ft. There are massive triangular cutwaters on the two midstream piers, both upstream and downstream, all four carried up to form pedestrian refuges. The downstream parapet on the south bank curls round to form a pen or keep, used to control sheep being washed in the river and it is this which gives the bridge its name. Some guide books claim that the practice still exists.

From Ashford the line of the bridge points towards the village of Sheldon, around which there is much evidence of lead mining. Mee suggests a seventeenth-century date.

Bakewell (Holme Bridge) (Group 1) 119: SK 215 690
DOE Listed Bridge. Derbyshire County number 17.

The well-known Holme Bridge spans the River Wye upstream of the town centre in five segmental plus two semi-circular arches. The overall length is about 102ft and the bridge is 48in wide between 27in parapets. Triangular cutwaters, both upstream and downstream, protect all the piers and are carried upwards to form pedestrian refuges. The roadway is flagged and the parapet coping stones are stapled. Chiselled into the coping stone of one of the downstream refuges is a slot which once held a cross, thought to have been destroyed during the Civil War.

The "official" notice alongside the bridge implies that it was built away from the town centre in order to avoid the tolls payable in the market area. The bridge was built in 1664 and although Jervoise describes it as a footbridge leading to Holme Hall, the notice confirms it to be a packhorse bridge on a route to the north. The word "Holme" derives from the Old Norse word *holmr* which translates as "water meadow". (See also Drigg in Cumbria.)

Barber Booth (Group 1) 110: SK 088 861
This is commonly known as "Jacob's Ladder" Bridge or is sometimes called Youngate Bridge. It spans the River Noe about 2 miles west of Edale and is on an alternative route of the Pennine Way long distance footpath at the foot of the steep path known as Jacob's

Ladder. In the second half of the eighteenth century a man called Jacob Marshall occupied the now ruined Edale Head Farm just above the bridge. He kept a small enclosure for packhorses to graze and is credited with beginning the steep direct path up the hillside to give the packmasters a respite whilst their horses took the longer zigzag route, hence the Jacob's Ladder name.

The bridge is on an important packhorse route connecting Yorkshire and Cheshire, parts of which are called "Monk's Road" and the probably medieval "Edale Cross" marks the top of the steep climb from the bridge. Part of this route is described in more detail under "Hayfield".

Both the path and the bridge were restored by the National Trust and the Manpower Services Commission 1987 as a joint contribution to the European Year of the Environment. The bridge has a span of 12ft and is 26in wide between parapets which are now 36in high.

Beresford Dale (Wolfscote) (Group 3) 119: SK 131 584

I have found only one (anonymous) reference to this as a packhorse bridge and judging by the criteria outlined in the preface, it isn't one. However, the construction is so unusual that it deserves a mention.

Two massive stone piers shaped like upturned boats and constructed without mortar are founded in the river (the Dove). The three resultant gaps are spanned by twin baulks of timber across which are laid sandstone paving slabs. The total span is about 42ft and the width of the slabs (the roadway width) is 32in. There are wooden handrails either side and as in many footbridges they are wider apart at the top than the bottom.

Jervoise mentions that the Staffordshire Sessions in 1709 ordered a gratuity towards converting "a wood horsebridge into a stone bridge for horses" at some point across the River Dove (exact location unknown) between Alstonfield and Hartington. The Dove valley is so narrow and steep sided that this small area at the junction of Beresford Dale and Wolfscote Dale, being much less steep is an obvious choice for a crossing place. Also, a line of mostly unpaved lanes and bridleways connect Hartington and Alstonfield and on the 1864 edition of the one-inch OS map, they are shown crossing the river at this point.

In the absence of any Dartmoor granite, or tough millstone grit nearby, the three spans, each about 14ft, seem too long for a stone clapper bridge and there is no obvious evidence that the existing drystone piers or the abutments supported an arched bridge.

I speculate that the 1709 bridge might have been on this site and might have been of the same construction as the present bridge - that is stone piers, wooden beams and a stone slab roadway.

Edale (Group 1) 110: SK 123 860

This exceptionally narrow bridge (28in between parapets), crosses Grindsbrook just east of the Nag's Head public house in the hamlet of Grindsbrook Booth a few hundred yards from Edale. The span is about 18ft and the parapets which are capped with some triangular and some square copings, are 30in high. The roadway has been repaired using tarmacadam and the downstream parapet has a distinct bulge.

The bridge is on an important long distance packhorse route between Cheshire and Yorkshire, a 15 mile section of which is described under "Hayfield".

Edale was a minor junction of packhorse ways. One route heads for Castleton via Hollins Cross, another south to the skyline notch of Mam Nick (Grid Ref: SK 125 834), where yet another east-west route is joined.

Edale, the southern end of the Pennine Way long distance footpath, has a large car park and a National Park Visitor Centre. It is much busier now than it ever was at the height of packhorse activity.

Goyts Bridge (Group 2) 119: SK 013 733

This bridge was rebuilt in its present position by Stockport and District Water Board in August 1965 to prevent it being drowned in the newly constructed Errwood reservoir. It now crosses the River Goyt just upstream of the reservoir.

Although called Goyts Bridge, the original of that name had a roadway 8ft 6in wide; this bridge crossed Wildmoorstone brook about 100yds from Goyts Bridge at Grid Ref: SK 014 749. Buildings adjacent to the original Goyts Bridge before their demolition were dated 1762 and it is probable that both bridges date from around that time.

The span of the rebuilt bridge is 12ft and the width 41in between 20in parapets. Both Goyts Bridge and this bridge were on an old salt route leading from the Cheshire "Wiches" to Buxton via Macclesfield, other clues on this route being Saltersford Hall about 2 miles west and a Saltersford Bridge over the River Dane about 1 mile north-east of Holmes Chapel in Cheshire.

Haddon Hall (Group 3) 119: SK 235 663

This is one of a handful of bridges which I visited but failed to reach. No rights of way or paths lead to it and since the bridge is close underneath the garden walls of Haddon Hall itself and a paying visitor would have to climb a high wire fence under the gaze of attendants, I didn't attempt it. It can be seen from the gardens and looks a nice bridge. Obviously narrow (Jervoise says less than 3ft), with parapets and formed of two arches which are so obtusely pointed as to be almost segmental.

The bridge figures in the well known legend of Dorothy Vernon, daughter of the house, who, against her father's wishes, eloped with John Manners, son of the Earl of Rutland some time in the first half of the sixteenth century, using the bridge as an escape route. The bridge is known as "Dorothy Vernon's Bridge" and assuming its use during the elopement to be fact, then Dorothy Vernon's bridge pre-dates the main bridge to the Hall which was probably built in 1663.

The packhorse bridge does not look medieval and in any case it seems improbable that access to an important Hall such as Haddon would be limited by a bridge less than three feet wide. Many sources describe Dorothy Vernon's Bridge as a packhorse bridge, but perhaps Jervoise is right when he refers to it as a footbridge.

Hayfield (Group 1) 110: SK 050 870

This tiny bridge spans the River Sett about three quarters of a mile upstream from Hayfield village and about the same distance below the Kinder reservoir dam. Known locally as "Bowden Bridge" or "Old Roman Bridge", the span is 20ft and the width 37in. There are rudimentary kerbs 8in high.

For anyone who relishes a 15 mile walk, very little of it on metalled road, a packhorse route can be followed from The Packhorse

The tiny Bowden Bridge, near Hayfield, Derbyshire

public house in Hayfield, which takes in three packhorse bridges plus a medieval cross and a Jaggers Clough. From Hayfield the route is via this bridge, Coldwell Clough, Edale Cross (Grid Ref: SK 077 861), Barber Booth packhorse bridge (q.v.), the packhorse bridge at Edale (q.v.) Nether Booth, Jaggers Clough (Grid Ref: SK 154 872) to the restored Hope Cross where the route crosses the line of the Roman road between Glossop (MELANDRA) and Brough (NAVIO). The route extends beyond Hayfield towards Charlesworth where it is called "Monk's Road", and beyond Hope Cross towards Sheffield.

Hollinsclough (Group 1) 119: SK 062 668

This semi-circular arch springs from high above river level and crosses the River Dove about a quarter of a mile north-west of the hamlet of Hollinsclough. The span is 15ft and the width 50in between 18in parapets. The copings have been rendered with cement and the roadway covered with tarmacadam.

The bridge - sometimes called "Hopping packhorse bridge" - is on an old packhorse route connecting Longnor with Hollinsclough and Booth, thence across Axe Edge towards the Goyt Valley where perhaps it joined the salt road crossing Goyts Bridge (q.v.).

Milldale (Group 1) 119: SK 139 546
This is "Viator's Bridge", made famous by Charles Cotton in volume II of *The Compleat Angler* and familiar these days to thousands of ramblers and school parties walking the length of Dovedale. There are two arches crossing the River Dove in a total span of 42ft. The width is 48in between 37in parapets which may have been heightened, as the quotation from *The Compleat Angler* implies.

VIATOR	What's here? the sign of a bridge? Do you use to travel with wheelbarrows in this country?
PISCATOR	Not that I ever saw sir. Why do you ask that question?
VIATOR	Because this bridge certainly was made for nothing else; why a mouse can hardly go over it: 'tis not two fingers broad.
PISCATOR	... but I have rid over the bridge many a dark night.
VIATOR	... I would not ride over it for a thousand pounds, nor fall off it for two; and yet I think I dare venture on foot, though if you were not to laugh at me, I should do it on all fours.

The Compleat Angler was first published in 1653, a clear implication that the bridge is older.

Sometimes called "Wheelbarrow Bridge", the bridge is owned by the National Trust and provides access from the south and west to the village of Alstonfield.

Raper Lodge (Group 1) 119: SK 215 652
This bridge crosses the River Lathkill in a delectable spot about half a mile north-east of the village of Youlgreave. Since it is known as Coalpit Bridge and the route which uses it Coalpit Lane; its original purpose is fairly obvious. Packhorse loads of coal from pits around Chesterfield were carried across the bridge to Youlgreave and beyond.

There are 3 arches with a total span of 36ft and the width is 58in between 26in parapets which look newer than the bridge. The overall hump of the bridge is noticeably shallow.

Slipperystones (Group 2) 110: SK 169 952

DOE Listed Bridge. Derbyshire County number 18.

This bridge is a rebuild with an interesting history. It was originally located in Derwent village, built by the Premonstratensian "White Canons" of Welbeck Abbey in Nottinghamshire in the seventeenth century. It was in great need of repair in 1682 and £100 was set aside for this. When Derwent village was submerged under Ladybower Reservoir in 1942, the bridge was dismantled after being photographed and the stones numbered. It was rebuilt in its present position in 1959 as a memorial to John Derry and the Sheffield branch of the CPRE.

The bridge presently crosses the River Derwent about half a mile upstream from the top of the northernmost of the Derwent reservoirs (Howden) in two arches, with a total span of 57ft. The width is 62in between 27in parapets. The midstream pier has upstream and downstream cutwaters which are continued upwards as pedestrian refuges and the roadway is paved with stone flags. A socket in the downstream refuge once held a cross.

Slipperystones site is on an old packhorse route called "Cut Gate" which led from Penistone and Langsett into the Woodlands valley via Rowley Farm. Ironically, the original crossing was a ford.

Three Shires Head (Group 2) 119: SK 009 686

This old bridge spans the River Dane at the junction of the counties of Cheshire, Derbyshire and Staffordshire, just above a pool which is known as "Pannier's pool". The bridge is probably seventeenth century but has been widened. Two distinct arches can be seen underneath, the one upstream (the more modern) 43in wide, the other 71in wide. Two iron through-bolts help tie the two arches together. The arch has a span of 27ft and is a shallow segment. The width is 8ft between 20in parapets, the copings of which are held together by iron staples and the roadway is modern concrete.

Three Shires Head is a favourite spot for ramblers and pony trekkers and is a junction of four old packhorse tracks. One track

Pony trekkers crossing the widened packhorse bridge at Three Shires Head,
where Cheshire, Derbyshire and Staffordshire join.

heads north for the Cat and Fiddle Inn via Danebower Hollow and
possibly onwards to Goyts Bridge (q.v.), one leads north-east towards
Buxton, one south-east to Flash and the other south towards
Gradbach.

The two square miles surrounding Three Shires Head is laced by
a network of tracks and walled lanes which probably joined the
many quarries and shallow coal pits scattered around the area. The
first edition of the 1in Ordnance Survey map of the area (1870-1896)
names several coal pits and collieries. There was even a Dane Bower
coal wharf just off the A54 near Grid Ref: SK 010 700.

Washgate Bridge (Group 1) 119: SK 052 674

This bridge is entered under its local name to avoid confusion with
"Hollinsclough", which has a packhorse bridge of its own. Washgate
bridge is a "classic". Isolated and approached by three rough and
deeply worn lanes (hollow ways), one of which is full of plump
bilberries in season and has lengths of well-preserved cobbled road
surface. The bridge crosses the River Dove (here the boundary
between Derbyshire and Staffordshire) and is only about 1 mile

north-west of Hollinsclough. The width is 55in between 10in kerbs and the span is about 24ft.

One lane on the Staffordshire side leads eventually to Longnor, the other towards Leek. In Derbyshire, the route leads towards Buxton.

Its simplicity and remote situation clinch this as the writer's favourite packhorse bridge in the Peak District.

Youlgreave (Group 1) 119: SK 215 641

This 36ft arch spans the River Bradford just below the church in the village of Youlgreave. The width is 45in between 28in parapets. The bridge is on the Youlgreave to Harthill bridle road and is known locally as the Bradford packhorse bridge.

There are two clapper bridges further upstream and although the one furthest upstream at Grid Ref: SK 209 640 looks modern, the supports of what may be a much older clapper bridge can still be seen in the river just downstream of it.

The packhorse bridge across the River Bradford at Youlgreave, Derbyshire.

GLOUCESTERSHIRE

Bibury (Group 3) 163: SP 115 066
This little bridge crosses the River Coln which here is a gin-clear stream full of big trout. Although having most of the attributes of a packhorse bridge, it was probably built as a footbridge to serve a row of seventeenth-century weaver's cottages.

The three small arches span about 30ft and the overall width (no parapets) is about 6ft 4in. The bridge has iron railings.

Lower Slaughter (Group 3) 163: SP 165 225
One of the many attractive small bridges which, although looking like a packhorse bridge is almost certainly a village footbridge.

The three small arches span about 18ft and the bridge is 34in wide between 6in kerbs. Sockets suggest that the bridge once had hand-rails.

Slad (Snow's Farm) (Group 1) 162: SO 887 081
A chance remark during a radio programme about the author Laurie Lee led to the hunt for this bridge. The best approach is down to the end of Stean Bridge Lane from the village war memorial. From there the footpath upstream can be followed.

The bridge, built of honey coloured cotswold stone, is tiny, with a span of 4 or 5ft across a small stream, not named on the map, but which flows into the River Frome. The crown of the arch is only 2ft above stream level and the overall width 50in (there are no parapets). The remaining abutments indicate that the bridge has been wider - perhaps twice as wide. There is damage on the upstream side (winter floods?), with arch stones having fallen into the stream. Perhaps the good people of Slad can persuade the powers-that-be to find the money to do a bit of restoration on it.

There is some indication of paving under the grass on the path up to the farm.

Local opinion has it that the bridge is on a "well-known packhorse track", used by pack trains carrying fleeces from Bisley to Gloucester.

HEREFORD AND WORCESTERSHIRE

Shell (Group 1) 150: SO 951 598
The hamlet of Shell is about four miles south-east of Droitwich and the packhorse bridge is almost certainly on one of the many salt routes which once radiated from the town.

From the Romans onwards the brine springs of Droitwich have been exploited by evaporation; not until the middle of the seventeenth century was rock salt mined. Apart from coastal evaporation, only the two inland areas of Droitwich and the Cheshire Wiches produced significant quantities. This restricted availability and the widespread need for salt to preserve meat and fish made long distance transport inescapable. Packhorse transport carrying this vital commodity must have been one of the very earliest organised distribution systems.

The bridge at Shell is small and crosses Bow Brook just downstream of a ford which carries the minor road. The total span is 21ft and the width 39in between kerbs 7in high. There are two semi-circular arches and the central pier has a low level triangular cutwater upstream. Jervoise assigns a seventeenth-century date.

LEICESTERSHIRE

Anstey (Village) (Group 1) 140: SK 552 084
DOE Listed Bridge. Leicestershire County number 35.
This is an unusual bridge which crosses the River Rothley on a green space a few hundred yards from the centre of the village. The bridge has five semi-circular arches spanning a total of 57ft. The width is 68in between 30in parapets. Each of the four piers has a triangular cutwater both upstream and downstream, all with a pedestrian refuge above (a generous eight in total). The roadway is cobbled, the parapets have brick copings and in the rough masonry of the bridge are massive stones of what to the untutored eye looks like granite, though the outcrops of pre-Cambrian rocks on Charnwood Forest, a few miles to the north-west, are a more likely source.

The bridge is on an old trackway called Anstey Lane (Anstey derives from an Old English word meaning "narrow footpath") which leads from Leicester Abbey north-westwards towards Charnwood and perhaps connected the Abbey with the Priories of

The 5-arch packhorse bridge in Anstey village near Leicester.

Ulverscroft, Charley and Grace Dieu. Other speculations suggest that Anstey Lane and the bridge site are on the line of the VIA DEVANA, the Roman road (not named until the eighteenth century) linking Colchester and Chester.

There is disagreement concerning the date. W.G.Hoskins suggests *circa* 1500, Jervoise late seventeenth century.

Anstey (Group 3) 140: SK 557 090

Less than half a mile downstream of the Anstey packhorse bridge (q.v.) is an old two-arch bridge with a massive triangular upstream cutwater which is carried up to form a pedestrian refuge. The span is about 20ft and the width 9ft 6in between 32in parapets.

The bridge is on an unsurfaced lane leading to Astill Lodge and is known as "King William's Bridge". Jervoise speculates that if the King is William III, then the bridge must date from the late seventeenth century.

Aylestone (Group 1) 140: SK 568 010
DOE Listed Bridge. Leicestershire County number 10.
This bridge across the River Soar is at the western end of Marsden Lane on the southern outskirts of Leicester. There are eight main and three subsidiary arches spanning a total of about 150ft. The width is 58in between 36in parapets which are topped by triangular copings. The piers have upstream cutwaters, two of which are carried up to form pedestrian refuges. One downstream pier also has a cutwater and refuge. The bridge is massively constructed and is probably fifteenth century, though some arches have been lined with brick at a later date. Though sometimes referred to as a packhorse bridge, Jervoise describes it as a footbridge and since it is now embedded in Leicester's southern suburbs, no packhorse route suggests itself. It is now a feature in Leicester City's 2,400 acre Riverside Park, winner of an Europa Nostra diploma of merit in 1990.

Bottesford (Group 3) 130: SK 807 391
According to Jervoise, this bridge was built between 1581 and 1620 by one Samuel Fleming, who was rector of Bottesford. He built it because he once nearly drowned whilst crossing the river (the Devon) and wished others to avoid the risk. (See also Durham - Deepdale Bridge.)

There are two segmental arches, one with three ribs and one with four, spanning a total of about 20ft. The present width is just over 6ft between 40in parapets with stapled copings and the central pier has an upstream cutwater.

"Fleming's Bridge" as it is known, gives access by foot to the church.

Enderby (Group 2) 140: SP 552 985
DOE Listed Bridge. Leicester County number 48.
This old bridge is one of the few which nowadays cross only dry land, the river having been diverted. It is also one of the few which the writer visited but failed to reach (due in this case to barbed wire). The following details are taken from Jervoise. The river is the Soar and the bridge was on the old road between Blaby and Enderby. There are two pointed arches with a total span of 36ft and an overall width of 6ft.

Medbourne (Group 1) 141: SP 799 931
DOE Listed Bridge. Leicester County number 42.
This four-arch bridge, three of which now span dry land, crosses Medbourne brook in the middle of the village and leads almost directly into the churchyard of the twelfth-century church. The arches are semi-circular, built from dressed sandstone and are all different sizes. The total span is 66ft and the overall width 72in. There are no parapets. There are three massive triangular cutwaters upstream and a ford downstream. The bridge is well maintained and has a modern brick-paved roadway and wooden handrails. A notice on the bridge claims it to be thirteenth century and implies that an earlier crossing to the south was on the line of the Roman road between Leicester and Huntingdon. In Leicestershire, this road is known as the Gartree Road (a Saxon name), and is the so-called VIA DEVANA, the Roman road linking Colchester and Chester. This puts it on the same line as Anstey Lane (see Anstey) and the line of the route, irrespective of its use by packhorses, is of great age.

Rearsby (Group 1) 129: SK 651 145
DOE Listed Bridge. Leicester County number 43.
According to Pevsner, this bridge is medieval, but above the keystone of the sixth arch (counting from the north-west) on the upstream side is a rectangular stone inscribed RH 1714. The Rearsby Local History Society interpret this as the initials of the builder (Robert Harrison) and the building date of the bridge in its present form.

There are seven semi-circular arches, some of which are slightly deformed and the seventh arch (counting from the north-west) is now almost buried. The total span is 66ft and the width 63in between 21in parapets. Each pier has a triangular cutwater on the upstream side only. The bridge is in the village on a line downhill from the church and crosses a small stream which joins the River Wreake. There seems no evidence of a long distance packhorse road, perhaps the bridge was associated with Rearsby Mill on the River Wreake about half a mile to the north-west.

Thurcaston (Group 1) 129: SK 566 109 and 560 106
There are three interesting bridges at Thurcaston.

Near to a timber-framed house called "Latymer House" (Bishop Latimer's House), which Pevsner dates to 1568, is a public footpath to Cropston, a few hundred yards down which are two bridges. The first no longer spans a water course, but a swampy-looking pool bordered by a barbed wire fence to keep stock out. The meadows here are low-lying, so perhaps the bridge serves its original purpose when the River Rothley is in flood. The bridge is without parapets, is 9ft 9in wide overall and the almost pointed arch spans about 7ft. The roadway is of grass-grown cobbles and continues as a causeway for about 55yds to the second bridge.

This is called locally "Sandham Bridge", the two semi-circular arches span about 24ft and the width between 26in parapets, is 68in. The central pier has triangular cutwaters both upstream and downstream and both are carried up to form pedestrian refuges. The parapets, which are topped by rounded copings of cement rendering and carry large patches of the lichen *xanthoria*, are stepped out at the Thurcaston end: that upstream about 3ft, the downstream about 1ft. There is a slight flare to the parapets at the other end. The roadway is of neatly set cobbles.

About $1/4$ mile upstream from Sandham Bridge is another bridge of almost identical construction. The two semi-circular arches span about 33ft and the bridge is 63in wide between 27in parapets. Again there are triangular cutwaters both upstream and downstream. The bridge is reached by a footpath starting just south of Thurcaston church.

The purpose of the routes served by these two bridges, both now carrying footpaths between Thurcaston and Cropston, was probably for local traffic, possibly joining Anstey Lane heading north-west into Charnwood Forest (see Anstey).

Jervoise makes a brief reference to Sandham Bridge but does not estimate its age. All three bridges are similar in appearance to that at Rearsby (q.v.), so I will hazard a guess that they were built in the late seventeenth or early eighteenth century.

LINCOLNSHIRE

Scredington (Group 2) 130: TF 098 409
DOE Listed Bridge. Lincoln County number 89.

This is a well-preserved but unused bridge alongside a minor road which crosses North Beck. The beck is culverted under the road which dips so as to revert to a ford under flood conditions. There are two segmental arches with a total span of 20ft and the overall width is 9ft (no parapets). The roadway has a cobbled surface. A building date as early as 1250 is sometimes claimed, which, if accurate, makes it one of the earliest in the country. As with the other two medieval packhorse bridges which remain in Lincolnshire, the likelihood is that they were located on routes leading to the wool port of Boston (see also Utterby and West Rasen).

Utterby (Group 1) 113: TF 305 932
DOE Listed Bridge. Lincoln County number 29.
This tiny arch crosses an unnamed stream which flows into the Louth Navigation (a canal which was opened in 1770) and is alongside a modern road bridge just downhill from the church. The span is a mere 9ft and the overall width (no parapets) is 62in. The roadway is cobbled. All authorities agree that this bridge is "ancient" and the three chamfered ribs under the arch support this. It is probably of the same period as the packhorse bridge at West Rasen (q.v.).

Less than 1 mile north of Utterby is the line of a Roman Road leading from Grainthorpe to Lincoln. This same route, called variously Salters Lane and Saltergate, was used in the Middle Ages for the transport of salt inland from evaporating pans on the coast. It is a pleasant conjecture, though without evidence, that Utterby packhorse bridge might have been used to ensure that the salt stayed dry.

A more likely supposition is that this diminutive bridge once supported strings of packhorses carrying wool. Utterby village is only 3 or 4 miles north of the market town of Louth, which boasts four inns with wool-related names (Packhorse, Woolpack, Crown and Woolpack, and Golden Fleece). Also there was a Cistercian Monastery at Louth Park, just east of the present town. Cistercian establishments were amongst the foremost wool producers in the early Middle Ages.

West Rasen (Group 1) 121: TF 062 893
DOE Listed Bridge. Lincoln County number 9.
This handsome fourteenth-century bridge crosses the River Rase alongside a modern road bridge in the village. There are three segmental arches, one of which now crosses dry land. The total span is 42ft and the bridge is 54in wide between 24in parapets. The two piers have triangular cutwaters upstream and the roadway is cobbled. Each arch has three chamfered ribs and a string course. Old maps show an almost continuous line of paths between the pre-conquest settlements of Market Rasen and Kirton-in-Lindsay via West Rasen, so the crossing point may be very old. The bridge could also be on a medieval route connecting Newstead Priory (10 miles to the north) with Boston, which at the end of the thirteenth century was England's busiest port for the export of raw wool.

Jervoise seems to have confused this bridge (in name only) with Bishop's Bridge which is 2 miles to the north-west.

NORFOLK

Houghton St Giles (Group 3) 132: TF 928 358
This 10ft span semi-circular arch crosses the River Stiffkey just downstream of Houghton St Giles village and less than one mile upstream from Walsingham (q.v.).

The construction is mainly of brick, but the downstream keystone is stone and is dated 1709. There are also two stone voussoirs near the arch haunches which are initialled, GL (right-hand looking upstream), and TF (left-hand looking upstream). There are no parapets, nor signs of there ever having been and the overall width is 80in. The bridge is disfigured by white-painted iron railings dated 1912.

Whether this is a packhorse bridge or a footbridge, as Jervoise suggests, is a problem posed by many other small bridges. In this case, the site is sufficiently far removed from the village to suggest a bridge built for access to farmland.

Walsingham (Group 1) 132: TF 936 367
This may or may not be a packhorse bridge. Coloured postcards on sale in Walsingham village suggest that it is, but the Norfolk

Museums Service think that an alternative use may have been to carry the Abbey precinct wall over the River Stiffkey.

Whatever its purpose, the bridge crosses the river within the grounds of Walsingham Abbey by four small arches (the two inner semi-circular, the two outer segmental). The total span is about 33ft and the overall width (no parapets) 72in. Two lateral through-bolts strengthen the masonry which is sandstone and there are half-height triangular cutwaters upstream. The upstream face of the bridge contains fragments of stone with traces of ecclesiastical-style carving. The pathway is partly grass-grown.

The Abbey (really a Priory), was founded in the middle of the twelfth century to serve the famous shrine to the Virgin Mary and the pilgrims who visited it. It seems possible that the bridge was, at some time after the Dissolution in 1538, repaired using stones from the ruins.

NORTHAMPTONSHIRE

Charwelton (Group 1) 152: SP 535 560
DOE Listed Bridge. Northants County number 6.
This thirteenth or fourteenth-century bridge is one of the oldest packhorse bridges in England. It crosses the infant River Cherwell immediately alongside the A361 in the village of Charwelton. The width is 55in between 13in parapets and the total span about 21ft. These dimensions, which are my own measurements, do not agree with those quoted by Jervoise (span 16ft and width 3ft). The bridge's two pointed arches indicate a probable fourteenth-century date.

One of the pre-conquest salt routes heading east from Droitwich and crossing the River Avon at Stratford, has been proposed as reaching Priors Marston which is only four miles west of Charwelton. The same line (on maps at least) continues south-east towards Towcester. Both before and after the Norman conquest, salt was carried long distances. It is possible that packhorse trains used this medieval bridge whilst carrying salt from Droitwich towards the east.

The medieval packhorse bridge at Charwelton in Northamptonshire.

SHROPSHIRE

Rushbury (Group 1) 138: SO 513 916
DOE Listed Bridge. Shropshire County number 30.
Guarded in summer by nettles, hogweed and waist-high grass and hidden in trees, this small parapetless arch across the Lakehouse Brook has to be searched for.

The minor road southwards from Rushbury church dips to cross the brook before climbing "Roman Bank" on to Wenlock Edge. The packhorse bridge is 100yds or so downstream of the road bridge and is best reached from the north.

Built using small, undressed stones, the bridge spans about 12ft and is 60in wide overall. The banks of the brook at the bridge site are precipitous and 2 or 3ft high, which is probably the reason for the substantial abutments.

From Cardington to the north, there is an almost continuous line of lanes and bridleways via Rushbury and the bridge, across Wenlock Edge to Diddlebury, thence by minor roads to Ludlow. This may well have been part of a packhorse road connecting Shrewsbury and Ludlow.

128

Beckford Bridge, Devon
Fifehead Neville Bridge, Dorset

Preston Bridge, Dorset crosses the River Jordan! *(A.Menarry)*

Avebury Bridge, Wiltshire *(A.Menarry)*

The longest packhorse bridge in England. At Great Haywood in Staffordshire.

STAFFORDSHIRE

Great Haywood (Group 1) 127: SJ 995 226
DOE Listed Bridge. Staffs County number 19.
This bridge, called variously Essex Bridge or Shugborough Bridge, has several distinctions. It is the only packhorse bridge across the River Trent; at 100yds it is the longest packhorse bridge in the country and it has more arches - fourteen - than any other packhorse bridge.

Jervoise, quoting Chetwyn writing in 1679, says that Shutborough (Shugborough) was "formerly joined to Haywood by a wooden bridge, which being ruinous, was in ye last age rebuilt with stone and contains 43 arches, at ye end of which stood ye Bishop's Palace". Jervoise speculates that the missing arches possibly formed a causeway, though there is no evidence of a causeway now. Pevsner assigns a sixteenth-century date.

Immediately upstream, two arms of the River Sow join the Trent which, just before the confluence, flows over a shallow weir.

The arches are segmental and each of the thirteen piers has

129

triangular cutwaters both upstream and downstream and each is carried up to form a pedestrian refuge (26 in all). Four arches at each end now span dry land at normal river flow and the last two arches away from Great Haywood village are set at an angle to the main crossing.

The bridge is 53in wide between 21in high parapets topped by chamfered copings, some of which, in the angles of the refuges, are single blocks forming the angle and some of which are stapled together with the normal form of leaded-in iron strap. The coping stones on the upstream parapet over the fifth arch from Great Haywood village are stapled by modern stainless steel pins about the diameter of a pencil. The roadway is tarmac and the bridge well used by dogs, pedestrians with and without prams and cyclists, all making use of Shugborough Park.

SUFFOLK

Cavenham (Group 2) 155: TL 766 695
DOE Listed Bridge. Suffolk County number 171.
This small, derelict, segmental arch crosses a tributary of the River Lark alongside a more modern bridge carrying a minor road, the line of which follows the "Icknield Way". It is thus on a route of very great antiquity.

The span of the bridge is about 10ft and the overall width 90in. There are now no parapets, but traces remain, the width between them about 60in.

The arch is an oddity in that when viewed from downstream, the bridge appears to consist solely of three brick arch-rings. Mr A.A.Watkins, in a Suffolk Institute of Archeology and Natural History paper, suggests that from the type of brick used, the age of the bridge cannot be later than the fifteenth century.

I have been unable to unearth any evidence (other than width if there were parapets), that this bridge was used by packhorses. A route from Newmarket in the south-west and following the Icknield Way is evident on the map. Beyond the River Lark, about two miles north-east, there is no obvious continuation.

Moulton (Group 1) 154: TL 698 654
DOE Listed Bridge. Suffolk County number 17.
Many books describing transport through the ages devote a chapter to the packhorse. If an illustration of a packhorse bridge is included, there is a good chance that Moulton will be chosen. It is a superb example, crossing the River Kennet by four pointed arches in a total span of about 63ft and is 60in wide between 26in parapets. The arches themselves are brick, the spandrels and parapets, flint; the only packhorse bridge in my knowledge to use these two materials in this way. There are triangular cutwaters both upstream and down, but no pedestrian refuges.

The notice alongside the bridge quotes an early fifteenth-century date, but since Moulton was a market town at the end of the thirteenth century, it is possible there was a bridge on the site to serve local market traffic before the present structure.

The notice suggests that the bridge served a packhorse road joining Bury St Edmunds and Cambridge.

WARWICKSHIRE

Tidmington (Group 1) 151: SP 245 375
DOE Listed Bridge. Warwickshire County number 65.
From the hamlet of Tidmington, a cul-de-sac farm road leads westwards at the end of which are the deserted villages of Ditchford Frary and Lower Ditchford. About half way along this lane, a very muddy and cow-churned field path gives access to the bridge.

From older maps, a possible packhorse route can be suggested between Moreton in Marsh and Shipston-on-Stour, via Todenham and the bridge and the best clue as to purpose is perhaps in place-names. The village of "Salford" about six miles towards the south-east and a "Salter's Barn" about the same distance north, suggest a connection with saltways from Droitwich. Its position at the northern end of the Cotswolds hints at wool as another packhorse cargo.

The bridge crosses the River Stour in two semi-circular arches about one mile downstream of the confluence of Paddle Brook and Knee Brook. All the arch-stones are squared, dressed and well-fitted, the wing walls of rubble. There is a massive cutwater to roadway level upstream and the total span is 33ft. The bridge is without parapets and the overall width is 72in. The roadway surface is concrete and the bridge is decorated by amateurish railings made up of galvanised pipes, wire and binder twine, no doubt to protect the stock which obviously use the bridge. An unused ford lies just downstream.

WEST MIDLANDS

Hampton-in-Arden (Group 2) 139: SP 213 801
DOE Listed Bridge. West Midlands County number 35.
This interesting bridge crosses the River Blythe less than a mile south-east of Hampton-in-Arden church. The surroundings have been made into a pleasant picnic spot with a few tables and parking for a dozen or so cars and a modern raised "causeway" of wooden duckboards leads to a bridge at either end.

The bridge has five arches, the three nearest the village are pointed in shape and built of smooth red sandstone. The other two are segmental and brick-built. The total span is about 70ft and the width overall (there are no parapets) about 75in. There is an iron handrail on the upstream side.

The river at normal flow is confined to the three pointed arches and these pier footings are reinforced with concrete. Jervoise writes that the ford alongside is "dangerous", so perhaps in flood the two brick arches also carry some flow.

Upstream, all four piers have triangular cutwaters up to road level. The downstream cutwaters numbers 1, 2, and 4 (counting from the village) are square buttresses with chamfered tops like lean-to roofs. The other (number 3) is stepped, somewhat like a miniature ziggurat. The roadway is edged with blue brick over the brick arches, sandstone over the others. The whole strongly suggests two phases of construction.

Jervoise suggests that the bridge is on a track between Hampton in Arden and Berkswell, but a main railway line and the A452 trunk road cross this and make it difficult to follow. Older maps suggest a line from Kenilworth to Birmingham.

THE BRIDGES OF THE
SOUTHERN REGION

AVON
Chew Stoke
Chewton Keynsham
Wickwar (Roman Arch)

CORNWALL
Launceston

DEVON
Beckford Bridge
Fingle Bridge
Lizwell Bridge
Sidford

DORSET
Fifehead Neville
Holwell
Preston
Rampisham
Sharford
Sturminster Marshall
Tarrant Monkton

SOMERSET
Allerford
Alford
Bruton
Bury Bridge
Dunster
Horner
Ilchester
Queen Camel
Rode
Tellisford
Watchet
West Luccombe
Winsford

WILTSHIRE
Avebury
Coombe Bissett
Monkton/Holt

INTRODUCTION TO THE SOUTHERN REGION

This Southern Region of the Guide illustrates very well the way in which the relics of packhorse transport have largely disappeared in the east whilst surviving in the west. The barren highlands of Dartmoor and Exmoor are one reason, the rural, wool-growing nature of the old Kingdom of Wessex another. As important as these is the fact that since the ninth century, power and wealth have been concentrating in the south-east of England. These factors and the need to feed London's ever growing population have ensured that more wheels on better roads killed packhorse transport in the south-east sooner than elsewhere.

Since this guide is concerned with arched bridges, the famous clapper bridges of Dartmoor and further to the west in Cornwall, do not appear, although they were undoubtedly used by packhorses from the Middle Ages onwards. It has been said that in the area of high land, much of which is now the Dartmoor National Park, wheels were unknown until the nineteenth century. Assuming the truth of this, then everything for human subsistence, plus all the fodder and peat fuel for the upland farms must have been carried by pack animals.

There were two commodities which had a wider importance. Wool of course, and the old tracks across the moor such as the so-called "Abbot's Way" across the southern part of Dartmoor, or the wool route from north Devon to Ashburton via Chagford must have been important for this. Tin was the other. The tin exported from Cornwall before the Roman occupation came from further west than Dartmoor. It seems that tin mining on the moor started no earlier than the Middle Ages.

Near to the north Devon coast, where Exmoor falls dramatically to Porlock Bay and Blue Anchor Bay, there is an unusual concentration of packhorse bridges. Within about ten miles, Allerford, Horner, West Luccombe, Dunster and Watchet, each have a packhorse bridge. With one exception, all the authorities I have consulted agree the packhorse status of these five bridges. However, C.E.Kille, writing in the *Somerset Year Book* for 1933 ("Packhorse Bridges of West Somerset"), argues quite persuasively that they are footbridges, each leading to a particular church. There is no doubt that footbridges were built to make churchgoing easier (Bottesford in Leicestershire and Crosby Ravensworth in Cumbria are two examples) and these can easily be mistaken for packhorse bridges. The argument seems impossible to prove either way and I follow the consensus of written opinion.

Wool was again the most important trade cargo around Exmoor with market towns such as Dunster, like so many towns in England, owing their medieval prosperity to the manufacture of woollen cloth.

Most other remaining packhorse bridges in this Southern Region are confined to the counties of Somerset, Wiltshire and Dorset within the undefined boundaries of the old Kingdom of Wessex. The chalk uplands which straddle these three Counties were probably the first area in England to be settled. Here is the biggest concentration of Neolithic and later prehistoric monuments. From here radiate the first long distance tracks: the Ridgeway, heading eastwards to cross the Thames at Goring, then north-eastwards, perhaps to join the Icknield Way sweeping northwards into Norfolk; and the Harroway, heading for Dover in one direction and Cornwall in the other and used, so it is said, for the carriage of pre-Roman tin from Cornwall. These early routes were used as long ago as 1500BC by the international traders of the so-called "Wessex Culture", in whose tombs have been found ornaments from as far away as the Aegean. None of this proves the existence of trains of packhorses, but it confirms the long history of transport and trade.

In the early Middle Ages, the Cistercians were active in the production of wool. A few tithe barns survive from that time and the packhorse bridges at Bruton and Coombe Bissett are both Medieval.

Later the woollen and cloth industries were developed by the "Clothiers" of the south-west. Some of these were substantial capitalists, buying wool and arranging for home workers to spin or weave it before fulling and finishing the cloth ready for sale. Men such as John Winchcombe (Jack of Newbury) had, as early as the end of the fifteenth century moved away from a cottage based industry, anticipating factory production by concentrating his weavers (still hand-loom), together in a building. Another was William Stumpe who a little later, at the dissolution of the monasteries, bought Malmesbury Abbey from the Crown and filled all the outbuildings with looms. Packhorses were used extensively, locally to carry wool and yarn, and over long distances transporting the finished cloth to London and the ports.

Other goods of commercial significance included lead from the Mendips, tin from Cornwall and salt from evaporating pans on the Hampshire coast. Bristol and Southampton were the most important ports and when the pack trains carried these goods for export, they would return with imported cargoes of wine or oil. Exactly as would a present-day lorry owner, the packmaster tried to organise loads for his horses in both directions.

AVON

Chew Stoke (Group 3) 172: ST 558 618

This bridge is included as a Group 3 Bridge because I think there are sufficient grounds for doubt. The bridge is used by cars, is 7ft 6in wide and provides access to minor roads to the north-west of the B3114. The bridge does not look particularly old and in the absence of other evidence, could be classified as a small village bridge like many another.

However, a magazine article about Chew Stoke village, published early in 1993, describes it as a packhorse bridge, so some local opinion must consider it so. Interestingly, the same article calls the downstream ford "Irish Bridge", (see the entry for Fifehead Neville packhorse bridge in Dorset for a West Country explanation of an "Irish Bridge").

There are two deep segmental arches, spanning a total of about 18ft across an unnamed stream which joins the River Chew $\frac{1}{4}$ mile downstream of Chew Valley Lake. The width of 7ft 6in is between 36in parapets.

Chewton Keynsham (Group 2) 172: ST 655 664

DOE Listed Bridge. Avon County number 161.

This handsome bridge is slightly more than 13ft wide between parapets, so its status as a packhorse bridge seems dubious. The DOE's listing describes it as a packhorse bridge, so presumably there must be some supporting evidence.

The bridge presently connects Chewton Keynsham village with Uplands on the B3116, but with a little imagination it could provide an old crossing of the River Chew on a route between Bath and Bristol staying south and west of the more formidable barrier of the River Avon.

The bridge has two shallowly-pointed arches spanning a total of 42ft. The parapets are 42in high and the central pier has triangular cutwaters upstream and downstream. These are of ziggurat-like construction, rising vertically for a few feet, and are then stepped back. They reach a height roughly equal to the crown of the arch.

Wickwar (Roman Arch) (Group 1) 172: ST 731 882

The DOE list "Horse Bridge" in Wickwar (Avon County number 164), and any bridge with a name like that must alert the packhorse bridge enthusiast. Disappointment! Horse Bridge is something like 13ft wide and carries a secondary road.

However, about 1/2 mile upstream, across an upper reach of the Little Avon River (locally known as "The Break"), is a superb little two-arch bridge. The span is about 20ft and the width 60in between minimal 3in kerbs. The arches are semi-circular, raised to about 6ft above stream level at their centre and the central pier is protected by upstream and downstream triangular cutwaters up to the height of the arch-springing.

The local name of "Roman Arch" together with the cutwaters suggest that the bridge is old; but the maze of footpaths between Wickwar and the Cotswold escarpment to the east, make guessing at a packhorse route difficult.

The bridge is (June 1993) badly overgrown with small trees beginning to sprout from the stonework. It will be a pity if the bridge is allowed to decay to total ruin.

CORNWALL

Launceston (Group 2) 201: SX 327 851

DOE Listed Bridge. Cornwall County number 382.

There is a minor confusion concerning the name and location of this bridge. The DOE locates the packhorse bridge as above plus a St Thomas's Bridge at SX 329 851, whereas most other authorities refer to the packhorse bridge as St Thomas's Bridge. SX 329 851 is the Grid Reference for the road bridge which carries the A388 and which was built in the eighteenth century.

The packhorse bridge was probably built in the fifteenth century and crosses the River Kensey adjacent to St Thomas's church and is sometimes called "The Prior's Bridge" - there are the remains of a Priory nearby. Some guide books refer to the crossing point as the site of an old festival called the "Waterfeire" at which witches were ducked.

The bridge has a total length of about 70ft and the five arches are a mixture of deep and shallow segments with one span having a pointed arch. Both sides of the bridge are fitted with iron handrails (there are no parapets) the width between which is 54in. Slate slabs about 1ft wide outside the handrails make an overall width of about 6ft 6in. The pier between arches two and three (counting from the church) has cutwaters upstream and downstream carried up to roadway level and an old gas lamp is positioned on the upstream cutwater. The roadway between the railings is cobbled.

DEVON

Beckford Bridge (Group 1) 193: ST 266 015
DOE Listed Bridge. Devon County number 123.
The area of Devon bounded by the A30, the A35 and the A358 is criss-crossed by a tangled network of narrow lanes and bisected north-south by the River Yarty. Beckford bridge crosses the Yarty about two miles north-west of Axminster, the nearest village being Membury. The bridge is disused and just upstream of the present road crossing.

Beckford bridge spans 27ft and although the arch is segmental, the hump is high and the width 66in between parapets which are 26in high at the centre rising to 36in at the ends. The roadway is cobbled overlaid with concrete and, unusually, the parapets also are built using cobbles.

Henderson and Jervoise in *Old Devon Bridges* refer to Beckford bridge as "ancient", suggest that it has been widened upstream by 2ft, but do not suggest a packhorse route it may have served. Nor can I.

As befits a DOE "listed monument", the bridge is in good repair.

Fingle Bridge. (Group 2) 191: SX 743 899

DOE Listed Bridge. Devon County number 121.

Fingle Bridge crosses the River Teign at a favourite picnic spot, with a car park, forest walks and the Angler's Rest Inn alongside. Nearby are the Iron Age Hill Forts of Prestonbury Castle and Cranbrook Castle. The bridge is on a route linking Moretonhampstead and Drewsteighnton.

Granite-built in the sixteenth or seventeenth century, the bridge spans about 66ft and is 80in wide between parapets which are 28in high in mid-bridge, reducing to 17in at the ends. The parapets are topped by copings, some of which are square, some square with rounded tops.

The three arches are in the form of shallow segments springing from 4 to 6ft above river level and the two piers are protected by triangular cutwaters both upstream and downstream, all four of which are continued upwards to form pedestrian refuges. Two coping stones in each refuge on the southern pier are connected by leaded-in iron staples but there are no signs of any other stapling.

Lizwell Bridge (Group 2) 191: SX 712 741

DOE Listed Bridge. Devon County number 614.

Lizwell Bridge is one of the most difficult to reach and probably least visited bridges in this guide. It crosses the East Webburn River just upstream of its confluence with the West Webburn River. The obvious access from Buckland in the Moor was blocked (June 1991) by a commercial forestry company felling in the woods which cover the sides of this steep valley. Alternative access points at Ponsworthy or Cockingford suffer from parking problems, Devon lanes being as they are.

Any view of the bridge is badly obstructed by trees and the stonework supports a flourishing growth of ivy. The single segmental arch spans about 20ft and the overall width is about 10ft 6in (there are no parapets). The roadway is concrete which looks recently laid.

Local opinion asserts that "Lizwell Packbridge" once had parapets and that these were destroyed and the arch damaged (hence the concrete) by carting timber over the bridge. Whether this damage occurred during the 1939-45 war or more recently are two conflicting opinions.

The bridge would have been locally useful connecting Buckland in the Moor with Ponsworthy, perhaps to join the old route from Widdecombe via Dartmeet and Merrivale to Tavistock.

Sidford (Group 1) 192: SX 137 899
DOE Listed Bridge. Devon County number 273.
This is a strange bridge to include in Group 1, because it forms the footpath of the bridge carrying the main A3052 road. However, fixed to the parapet dividing the footpath from the road is a bronze plaque which explains all. It reads:

COUNTY OF DEVON
THIS PACKHORSE BRIDGE DATES FROM ABOUT THE YEAR 1100 A.D.
IT WAS PRESERVED IN ITS ORIGINAL SHAPE AND CONDITION
ON THE WIDENING OF THE BRIDGE IN 1930.
BRIAN S MILLER   CLERK
ANDREW WARREN   COUNTY SURVEYOR

The claim that the bridge dates from 1100 makes it the oldest packhorse bridge in the country. I wonder. That there was a bridge here in 1100 seems reasonable. That this is that bridge seems to me unlikely.

There are two arches with a span of about 60ft. The junction of the arches is more an island than a pier because two streams join under the road bridge. The River Sid flows from the north, and from the east, a stream which originates in Harcombe.

The footpath (and therefore presumably the original bridge) is 45in wide between parapets which are 31in high, capped with triangular coping stones. Across the road is a modern cul-de-sac named "Packhorse Close".

DORSET

Fifehead Neville (Group 1) 194: ST 772 111
DOE Listed Bridge. Dorset County number 12.
This bridge is decidedly odd. When a pointed arch is referred to, it normally means that each side of the arch is a radius and where they cross is the "point". Fifehead Neville is different because the sides of the arch are straight, the two arches are in fact triangles, even the

Probably the most unstable packhorse bridge still standing.
Fifehead Neville in Dorsetshire.

keystones are triangular. Two of the sides curve very slightly inwards, which would be the direction of collapse if too much load were applied.

Fortunately for the bridge, traffic on the minor road crosses the river, the Divelish, on concrete slightly upstream. The river is piped under the concrete at normal flows, the crossing becomes a ford when the river is in flood. This arrangement, according to A.J.Wallis in *Dorset Bridges*, is known in the county as an "Irish Bridge".

At Fifehead Neville, the two "arches" span about 15ft and are divided by a large triangular cutwater upstream. The width between 6in kerbs is 65in and wooden railings have been added. No route other than for local traffic is obvious.

The bridge is medieval and according to Jervoise, there is the site of a Roman villa nearby, though this is not marked on the 1:50,000 OS map. The bridge at nearby Sturminster Newton carries a "Transportation Plate" - see the chapter on Bridge History and Construction.

Holwell (Group 1) 194: ST 699 120

The village of Holwell and Bishops Caundle are joined by bridle ways and field paths via Holwell church; the packhorse bridge is located behind the church. A.J.Wallis in *Dorset Bridges* (1974) writes of three spans, one a slab, one of wrought iron and concrete and the third a stone arch. To the anger of at least one local resident, the slab and wrought iron spans were "knocked down by the council" and replaced by "an ugly wooden footbridge".

The stone arch remains and is undoubtedly part of a packhorse bridge. The shallow segment springs from four or five feet above stream level and spans about 12ft. The unaltered roadway is 62in wide between parapets which are 27in high. Upstream, a massive triangular masonry cutwater protects the arch pier and a similar pier and cutwater support the wooden footbridge at mid-length. Mr Wallis suggests that these piers are probably medieval.

Holwell packhorse bridge crosses the Caundle Brook and just upstream is Cornford Bridge which was built around 1480, and is DOE listed (Dorset County number 478). Cornford bridge has a span of more than 30ft, is very narrow with pedestrian refuges, but carries a minor road and was probably built as a "cart bridge".

Preston (Group 2) 194: SY 703 831

DOE Listed Bridge. Dorset County number 65.

This River Jordan (no more than 2 miles long) flows into Weymouth Bay through Preston. About 30yds upstream of the main A353 road bridge, in the garden of "Roman Bridge Cottage", "Roman Bridge" spans the Jordan. Perhaps it is so-called because the remains of a Roman villa were discovered nearby.

Referred to as a packhorse bridge in at least one publication, the DOE has it as a footbridge. Jervoise assigns a late medieval date; whether for packhorse or human foot, it is a most interesting bridge.

Spanning about 12ft, the overall width (no parapets) is 8ft 9in and the shallow segmental arch is quite rudimentary. There is a hint of a "flat" at roughly mid-arch which suggests eventual collapse - though as with other bridges, providing the abutments remain sound, they seem able to postpone "eventual collapse" for a few hundred years.

If Jervoise is right about the date, then packhorses most probably did use the bridge, though on what route it is hard to speculate.

Rampisham (Group 1) 194: ST 562 023

This old bridge is almost lost in a thicket of bushes and nettles near the village crossroads, though the footpaths leading to it have new wooden railings. There are three unequal, pointed arches and A.J.Wallis in *Dorset Bridges* observes that there can be a strange flow pattern: downstream through one arch, back upstream through the next and then downstream again through the third. On my visit (Oct 1992), water was passing only through the southern arch, so the stream was unable to perform. On the upstream side of the southernmost pier, a dividing wall several feet long protrudes into the stream; perhaps this influences the strange phenomenon observed by Mr Wallis.

A line of minor roads joining Powerstock to the south-west with Evershot to the north-east, suggests a possible packhorse route.

The bridge spans about 30ft across an un-named stream which joins the River Frome about two miles away at Sandhills. The concreted roadway is 63in wide between parapets which are built as double walls and are 32in high.

Just over a mile upstream from Rampisham, is Benville Bridge, which carries one of the "Transportation Plates" referred to in the chapter on Bridge History and Construction.

Sharford (Group 1) 195: SY 967 847

The tiny Corfe River, having cut through the Purbeck Hills just west of Corfe Castle, eventually reaches Poole Harbour. Sharford Bridge crosses the river as it wanders its way across Middlebere Heath.

On Saxton's County Map of Dorset (1575), at a position as near to Sharford Bridge as the crudity of the map allows judgement, is a bridge named "Sherford" with the words "bridge owre". It is thought that the bridge is on an ancient track joining Wareham and Newton which skirts the southern edges of Poole Harbour.

The bridge is very difficult to inspect because, except for the roadway, a jungle of nettles and ivy are trying to devour it. There are two arches which are crudely segmental and which span a total of about 18ft. The width is 60in between parapets which are 18in high and some of the coping stones are laid on end, making crude castellations.

Under the supervision of A.J.Wallis (see his *Dorset Bridges*),

Sharford Bridge was repaired in 1972 of the damage to the parapets and the two arches caused by ivy. It could be that the parapets were repaired again eight years later because the date 1/5/80 is inscribed in the cement pointing at the western end of the southern parapet. In Oct 1992, it looked as though the ivy was trying again to demolish the bridge.

Sturminster Marshall (Group 1) 195: ST 945 001

The main bridge in Sturminster Marshall is known as "Whitemill Bridge", was probably built in the twelfth century, attracts a DOE listing (Dorset number 73) and has a "Transportation Plate" - see the chapter on Bridge History and Construction. It is interesting also in that it was founded on vertical oak piles with an oak raft on top. During repairs in 1964 under the supervision of A.J.Wallis - (see his *Dorset Bridges*) the oak raft was replaced by one of concrete. See also Sutton in Bedfordshire for a packhorse bridge founded on wood, in that case, elm.

Much less well-known, slightly north-west of the village and crossing the River Winterborne, is a tiny packhorse bridge. The span is about 15ft and the width overall (no parapets) is 54in. The single arch is a very shallow (large diameter) segment and the bridge appears to comprise arch stones and kerbs only. Jervoise, in describing this bridge, remarks on the unusually long stone voussoirs and suggests an eighteenth-century date.

On my visit (Oct 1992) the river name "Winterborne" was apt. The stream bed was totally dry.

Tarrant Monkton (Group 1) 195: ST 945 091

The River Tarrant, which joins the Stour near Spetisbury and is only seven or eight miles long, gives its name to eight villages. Tarrant Monkton is the fourth village from the source. The bridge is in the village just downstream of a ford of a type which seems to be known in Dorset as an "Irish Bridge" - that is, pipes covered in concrete so that the road is dry except when the river is in flood - see also Fifehead Neville.

There are three arches which, although segmental, are nearly semi-circles. The bridge is only 41in wide between kerbs a few inches high and has a total span of about 20ft. Roadway and kerbs are both concrete covered. Modern white-painted wooden railings have been added.

The bridge is on a line of lanes, bridleways and field paths between Pimperne and Moor Crichel, but a more ambitious speculation suggests the possibility of a route between Blandford Forum and Salisbury keeping to low land south-east of Cranborne Chase.

SOMERSET

Allerford (Group 1) 181: SS 905 470

DOE Listed Bridge. Somerset County number 212.

This bridge has the distinction of appearing on the cover of the Landranger Map 181. It crosses an unnamed stream which joins Horner Water just before it flows into Porlock Bay.

The two shallow, segmental arches, span a total of about 27ft and spring from about a yard above normal stream level. The width between 23in parapets is 48in and there is a crude, low-level cutwater upstream. The roadway is neatly paved with small cobbles and the parapets are "roofed" with small flat stones laid like crazy paving. As befits a much photographed bridge, it is very well maintained. The upstream ford is still in use.

Viewed from upstream, the two arches are equal. From downstream, the north-east arch presents an odd skew shape, bending towards the north.

Although generally described as a packhorse bridge, it is difficult to propose any particular route which it might have served. As with many other bridges, its use was probably mainly for local traffic.

Alford (Bolter's Bridge) (Group 2) 183: ST 606 334

DOE Listed Bridge. Somerset County number 43.

By repute, this bridge, which crosses the River Alham $1/4$ mile before it joins the River Brue, was part of a packhorse road constructed by Monks to connect the Abbey at Glastonbury with the Priory at Bruton. Jervoise is less ambitious in suggesting a route connecting Hornblotton and Ansford. The bridge presently serves what seems to be a well-used bridleway. My own speculation is that Bolter's Bridge and Bruton Bridge (q.v.) were both on a packhorse wool-road leading from the west towards London.

A picture of Allerford packhorse bridge in Somerset graces the cover of Landranger map 181.

There are four unequal pointed arches, the westernmost almost completely blocked by mud and vegetation, the next one to it, dry. The total span is about 45ft with a few feet at either end of made-up approach. Bridge and approaches are all cobbled. There are no parapets and the overall width is 7ft. Upstream cutwaters protect the piers.

Bruton (Group 1) 183: ST 683 348
DOE Listed Bridge. Somerset County number 176.
This little bridge is one of the classic medieval (fifteenth-century) packhorse bridges. It crosses the River Brue (Celtic for "fast flowing) just downstream of Church Bridge which carries the main road. Both bridges were restored following damage by floods in July 1982.

Before the reformation, there was an Augustinian Priory in Bruton and it is probable that the packhorse bridge was built in part to serve this. As with many small towns in the West Country - see

This fifteenth-century packhorse bridge in the village of Bruton, Somerset.

also Dunster - early prosperity was based on wool, so the bridge might have served a west-east route. My speculation is that a wool-road from the west may have used Bolter's Bridge near Alford (q.v.) and Bruton Bridge on its way towards London.

The bridge construction is unusual. The main arch is of single chamfered blocks of dressed stone perhaps 2ft 6in wide. On either side of this are secondary arches of smaller chamfered blocks which are so far outwards from and above the main arch that they look like deliberate corbelling. The bridge is narrow, only 42in wide between parapets which are 33in high capped with stapled-together copings. The semi-circular arch spans 30ft and has a flagged roadway. The approaches are cobbled.

Immediately downstream is a ford and a set of stepping stones.

Bury Bridge (Group 2) 181: SS 945 274
DOE Listed Bridge. Somerset County number 11.
This interesting bridge which may have been built as early as the fourteenth century crosses the River Haddeo in the village of Bury just upstream of a ford which is still used. The bridge was once privately owned.

The bridge has four arches, the two to the south-east are pointed, the other two segmental, which suggests a reconstruction, although the stone has the same appearance throughout. The width between parapets which are high at 3ft 6in is about 6ft 6in and the total span is about 48ft. The roadway has a cobbled surface.

Each pier has a triangular cutwater upstream. On the downstream side, the cutwater on the pier between the segmental arches and that next to it are triangular but squared-off, more like buttresses than cutwaters, again suggesting that the two segmental arches are later reconstructions.

A line of minor lanes reach Bury Bridge from Wiveliscombe in the east and continue to Dulverton after crossing the River Exe by Hele Bridge.

Dunster (Gallox Bridge) (Group 1) 181 SS: 990 432
DOE Listed Bridge. Somerset County number 149.
Not only is this well-known medieval bridge DOE listed, it is a "Crown Property in the charge of the Secretary of State - DOE." As

Gallox Bridge at Dunster in Somerset.

with other old bridges, it is known by several names: "Roman Bridge", "Doddebridge", as well as "Gallox". There are two opinions on the derivation of "Gallox". One that it is derived from "Galloway", a breed of horse often used as a pack animal, the other (more likely in my view) because the gallows were once located just beyond the south-eastern end of the bridge.

The two shallow segmental arches with double arch-rings, which spring from stream level, span a total of about 25ft. The width is 46in between 24in parapets which have been cement rendered to a triangular shape. There is a low level cutwater upstream on the central pier. As a tourist attraction Gallox Bridge, which crosses the river Avill, is very well maintained.

Dunster was a prosperous medieval woollen town, so there seems little doubt that Gallox Bridge is a genuine packhorse bridge and supported many a horse load of wool into the town and cloth out again when the trade was at its peak.

Horner (Group 1) 181: SS 897 456
DOE Listed Bridge. Somerset County number 30.
One of three interesting old packhorse bridges within a mile or so of each other (the others are Allerford and West Luccombe). This one crosses Horner Water a few tens of yards from the official car park in Horner village.

The single arch is segmental and slightly distorted in a way that suggests a semi-ellipse. The span is about 18ft and the width 50in between 22in parapets which have cement rendered copings. The roadway is cobbled.

Although most authorities refer to this as a packhorse bridge (including the OS who name it using the script reserved for antiquities) it is possible that it served a path leading to a church (St Dubricius) - see the Introduction to the Southern Region. Pevsner writes that it is on the "Hacketty Way".

Illchester (Pill Bridge) (Group 1) 183: ST 500 234
DOE Listed Bridge. Somerset County number 186.
This three-arch bridge which crosses the River Yeo, is satisfying in that it is not too easy to reach, but worth the effort once there. The access lane just north of the village of Long Load on the B3165, has unwelcoming KEEP OUT notices. The approach from Ilchester is easier. At the southern end of the old High Street, Pillbridge Lane is navigable by car as far as a Gas Pressure Regulating Station. From there, a lane so overgrown as to be almost impassable leads to the bridge. The field alongside the lane makes for easier progress.

According to Jervoise, Pillbridge Lane is shown on Stukeley's plan of Ilchester (1723) as the "Langport Way". Jervoise also quotes Thomas Gerard (c 1633): "The river (Parrett) paseth under Pill Bridge whither are brought upp boates and crayes from Lamport and Bridgewater." *Note* that the Yeo joins the Parrett at Langport. Both Pevsner and Jervoise suggest that Pill Bridge was built in the early seventeenth century.

An item from the "Local Notes and Queries" in the *Somerset County Herald* of 3/11/1923 adds some confusion by claiming that Pill Bridge was built no later than 1400. However, the style and construction details of the bridge suggest that the seventeenth century is more probably correct. The same newspaper item goes on to suggest that there was a public house near the bridge until sometime in the nineteenth century and that coal was carried upriver to the bridge site until the coming of the railways.

The three arches are semi-circular, the central arch much the biggest and plenty high enough for small boats to pass through. The total span is about 54ft, the width 48in between 34in high parapets which are a massive 18in thick and which ascend to a shallow point at the centre of the bridge. These parapets are strengthened in an unusual way. Vertical iron straps down each face of the parapet are connected by leaded-in bolts across the top of the coping stones. In the centre, conventional leaded-in iron staples are used lengthways. On the downstream central coping stone, the outline of a large hand has been chiselled. It looks to be contemporary with the bridge.

Triangular cutwaters rise to half pier height both up and downstream and are then stepped back. The roadway, which enters the bridge via a marked flare in the parapets, is surfaced with cobbles, now grass-grown, and stock is excluded by wooden gates.

Pill Bridge is well built in dressed stone and is in good repair. A super bridge.

Queen Camel (Group 1) 183: ST 593 250
Jervoise makes reference to this bridge but does not suggest that it is a packhorse bridge. Local people are in no doubt that it is. Its position, well outside the village, crossing the River Cam to connect Blackwell Road with Laurel Lane, with no obvious present farm use, together with its obvious age and diminutive size, persuade me to agree with local opinion. Probably built for local use, it is on a line connecting Marston Magna with the Roman Foss Way in the vicinity of Keinton Mandeville.

The segmental arch, springing from about 3ft above stream level, spans about 15ft and the width between parapets is only 38in. The parapets are 30in high, capped by massive square section coping stones, stapled together. Built-up wing walls ensure that roadway and parapets remain horizontal. The roadway is of recent concrete.

Rode (Group 1) 183: ST 799 540
This three-arch bridge crosses the River Frome on a track (now a footpath) which joins Rode to a point east of Woolverton. The track

from Rode village downhill to the bridge is a classic "Holloway", now much overgrown.

The bridge has a total span of 63ft and is 48in wide between 40in parapets. The arches are semi-circular, the construction very rugged and the upstream cutwaters are massive.

Rode packhorse bridge is sometimes known as Scutt's Bridge.

Tellisford (Group 2) 173: ST 805 557

None of the obvious standard works (DOE, Jervoise, Pevsner) mention this bridge and I am indebted to a man who, when I asked for directions to the Rode packhorse bridge (q.v.) said, although we were in Rode, "I don't know of one in Rode, but there's one at Tellisford". It is true that Mee praises the prettiness of Tellisford village, but he makes no mention of the bridge either.

The road leading eastwards from the church is a dead end with a notice saying "No turning spaces", and as the river, the Frome, is approached the road becomes a narrow cobbled causeway which drops very steeply to the river. Although restored, I feel sure the cobbles were originally intended to help packhorses.

The bridge spans 66ft in three segmental arches which spring from about 4ft above normal stream level. The piers are protected by enormous triangular upstream cutwaters, balanced downstream by square buttresses. There are no parapets and the overall width of the cobbled roadway is 6ft 10in. Rather dilapidated white-painted wooden railings have been added.

Slightly west of the bridge and river, there is what looks like a mill race, with a weir upstream marked on the 1-inch OS map. Perhaps the bridge was built to serve a water-powered mill.

Watchet (Kentsford Bridge) (Group 1) 181: ST 058 426

Located at the end of the lane leading to Kentsford Farm from the B3190, this bridge is unique in that it retains a stone cross surmounting the low level upstream cutwater on the central pier.

The two arches are shallow segments with a total span of about 16ft and the width is 54in between 18in parapets.

The stream, the Washford River, divides near the bridge, the eastern branch having the engineered look of a mill leat, probably feeding the mill downstream which is still marked on modern maps.

Both the purpose and age of this bridge are difficult to establish. In the records of the Quarter Sessions for 1613, the inhabitants of the Parishes of Old Cleeve and St Decumens were excused from having to repair it, the responsibility lying elsewhere. It has been suggested that the bridge is even older than this, having been built before the Reformation. The proximity of Cleeve Abbey and the undamaged Bridge Cross lend credence to this.

West Luccombe (Group 1) 181: SS 898 461
DOE Listed Bridge. Somerset County number 29.

This small bridge crosses Horner Water less than $^1/_4$ mile downstream of the packhorse bridge at Horner. There are no hints on current maps of any purpose other than for farm access.

The bridge has a shallow pointed arch springing from high above normal river level and the eastern buttress is unusually massive. The span is about 15ft and the width between 30in parapets is 39in.

The roadway surface is cobbled and very rough and the copings are cement rendered.

Winsford (Group 1) 181: SS 905 351 and 905 349
DOE Listed Bridges. Somerset County number 38 and 197.

The village of Winsford boasts eight bridges, two of which are packhorse bridges, both of them on an old track joining Dulverton and Porlock and both of them DOE listed.

The larger of the two, which is at 905 351 and has County number 38, crosses the River Exe by two segmental arches spanning a total of about 30ft. The central pier has a low level cutwater upstream and the width of the cobbled roadway is 42in between 28in parapets which have rounded copings.

Jervoise quotes a report to the Quarter Sessions of 1628 to the effect that "Viccaridge Bridge" had been damaged "by the violence of the waters" and was "re-edified" at a cost of £5 13s 11d. The present structure called "Vicarage Bridge" is just downstream of the packhorse bridge, was built in 1835 and re-constructed in 1928. There is thus a clear implication that the Exe packhorse bridge is the original "Viccaridge Bridge".

The second packhorse bridge is a small segmental arch which

springs from water level and crosses the Winn Brook opposite the Royal Oak Hotel in the centre of the village. The span is about 9ft and the width 55in between parapets which vary in height between 26in and 14in. The roadway is cobbled.

WILTSHIRE

Avebury (Group 2) 173: SU 097 699

Jervoise mentions this bridge, describing it as "small" without giving an opinion as to whether it is a packhorse bridge or a footbridge. Located about $^1/_4$ mile west of the church, it crosses the upper River Kennet, described locally as "only a winterbourne" - i.e. it dries up in summer.

There are four semi-circular arches, with crowns about 3ft above stream level, but slightly uneven in height. Three of the arches have been repaired using brick, the westernmost arch is the only one with (original?) arch-stones. The total span is about 36ft and the width variable between 9ft and 9ft 6in overall. There are no parapets and the general appearance is spoiled by white-painted iron railings. The midstream piers have substantial "feet" but no cutwaters.

The pathway surface is tarmacadam, which covers possible evidence of a causeway. The area is boggy with outlying arches over small runnels about 10yds to the west and 60yds to the east of the main bridge.

I can find no direct evidence of packhorse use, but the bridge is small, old and leads into open country down an old lane named "Bray Street". It seems worthy of inclusion.

Coombe Bissett (Group 2) 184: SU 109 264

DOE Listed Bridge. Wiltshire County number 277

This medieval packhorse bridge is in the village, a short distance downstream of the road bridge. Both cross the River Ebble.

Jervoise speculates that the bridge may have been referred to by John Leland during his free-ranging travels between 1533 and 1539. Leland had been made King's Antiquary by Henry VIII, who had given him authority to search the Kingdom for curiosities. This he did comprehensively, reporting the journeys in his *Itinerary*. He was the first in a long sequence of travellers - Camden, Celia Fiennes,

The medieval packhorse bridge at Coombe Bissett in Wiiltshire.

Defoe - who have journeyed through Britain and reported on what they found, leaving an invaluable record of the countryside of their time. The process did not stop with Defoe of course; similar (though physically much easier) journeys are still being made.

The bridge has three shallow, pointed arches, spanning a total of 36ft. The overall width is 7ft 6in and although there are no parapets, wooden railings have been added. Low level cutwaters upstream are balanced by buttresses on the downsteam side of the piers.

It is difficult to see the most obviously medieval side of the bridge which is downstream; upstream, the arches have been repaired using brick. The roadway is tarmac and the bridge in good repair.

Monkton/Holt (Group 1) 173: ST 882 620
A couple of miles south-west of Melksham, the A3053 crosses above the railway line and just beyond the crossing, a field path heads south towards the River Avon and the packhorse bridge. Holt is the

Monkton/Holt - The packhorse bridge across the River Avon in Wiltshire.

nearest village and Monkton House the nearest habitation marked on the map, hence Monkton/Holt for the name of the bridge.

The packhorse bridge is a substantial but elegant construction in pale grey dressed stone and was probably on a track between the villages of Semington and Broughton Gifford, presently a line of field paths.

There are four arches, all semi-circular, but all of different sizes. Even the triangular cutwaters upstream on the piers are different sizes. The bridge spans a total of about 63ft and is 45in wide between parapets. These flare markedly at the northern end, ascend to a shallow point at the centre of the bridge, are 32in high and have stapled copings. There is a string course at their base.

Jervoise mentions the bridge, but does not estimate its age. My own unprofessional guess is that it is unlikely to be earlier than the eighteenth century.

The northern end of the bridge is fenced to exclude stock (a large herd of bullocks consumed by curiosity on my visit) and seems now to carry only a light traffic of ramblers.

SELECTED BIBLIOGRAPHY

GENERAL

The study of this subject would be impossible without Ordnance Survey maps. The 1:25,000 Outdoor Leisure maps and the 1in to 1 mile Tourist maps are especially useful, but as they cover only popular tourist areas, the maps used for reference in the Guide, are the 1:50,000 Landranger series. Interesting too, are the first editions of the 1in maps reprinted by David and Charles.

Addison Sir W. *The Old Roads of England.* Batsford, 1980.

Authors Various. *The Pelican History of England* (9 vols). Penguin.

Bonser K.J. *The Drovers.* Macmillan, 1970.

Bottomly F. *The Explorers Guide to the Abbeys, Monasteries and Churches of Great Britain.* Avenal Books, 1984.

Bridbury A.R. *England and the Salt Trade in the later Middle Ages.* Oxford University Press, 1955.

Crofts J. *Packhorse Wagon and Post.* Routledge and Kegan Paul, 1967.

de Mare E. *Bridges of Britain.* Batsford, 1954.

Defoe D. *A Tour through the whole Island of Great Britain.* Dent (Everyman), 1962.

Department of the Environment *Lists of Ancient Monuments in England* (3 vols). HMSO, 1977-8. The Regional boundaries used in this guide are the same as those used by the DOE. A word of warning; some of the Grid References quoted in these lists are incorrect. Also, not all packhorse bridges are described as such.

Fiennes C. (ed Morris C). *The Illustrated Journeys of Celia Fiennes.* Macdonald, 1982.

Hindley G. *A History of Roads.* Peter Davies, 1971.

Hoskins W.G. *The Making of the English Countryside.* Penguin, 1970.

Jervoise E. *The Ancient Bridges of England and Wales* (4 vols). Architectural Press, 1930-6. For anyone interested in old bridges, these volumes, describing over 5,000 bridges, are without equal. More than 50 of the bridges described in this Guide, were "discovered" in Jervoise. *Note* that the bridges described by Jervoise at Clunton (Shropshire) and Penistone (South Yorkshire) have now gone.

Johnson S.M. and Scott-Giles C.W. (eds.) *British Bridges - An Illustrated and Technical Record.* Organising Committee of the Public Works, Roads and Transport Congress, 1933.

Jusserand J.J. *English Wayfaring Life in the Middle Ages.* Cedric Chiver Ltd, 1970.

Lipson E. *A Short History of Wool and its Manufacture.* William Heinemann, 1953.

Margary I.D. *Roman Roads in Britain.* John Baker, 1973.

Mee A. *The Kings England* (by county). Hodder and Stoughton.

Millward R. and Robinson A. *Upland Britain.* David and Charles, 1980.

Pevsner N. *The Buildings of England* (by county). Penguin.

Power E. *The Wool Trade in English Medieval History.* Oxford University Press, 1941.

Smiles S. *Lives of the Engineers.* John Murray, 1862.

Taylor C. *Roads and Tracks of Britain.* Dent, 1979.

NORTHERN REGION

Boyes M. *Exploring the North York Moors.* Dalesman Books, 1976.

Collingwood W.G. "Packhorse Bridges". *Transactions of the Cumberland and Westmorland, Antiquarian and Archaeological Society,* vol XXVIII. 1928.

Cotton F.W. "Packhorse Bridges of the Lake Counties." *Naturalists Field Club and Photographic Society Proceedings.* 1963.

Drake M. and D. *Early Trackways in the South Pennines.* Pennine Heritage Network, 1981.

Hatcher J. *Richmondshire Architecture.* The author, 1990.

Hey D. *Packmen, Carriers and Packhorse Roads.* Leicester University Press, 1980.

Hey D. *Yorkshire from AD 1000.* Longman, 1986.

Hindle B.P. *Roads and Trackways of the Lake District.* Moorland Publishing, 1984.

Maxim J.L. "Packhorse and other Ancient Tracks in and around Rochdale." *Transactions of the Rochdale Literary and Scientific Society,* vols XV-XVI. 1923-8.

Orrell R. *Saddle Tramp in the Lake District.* Robert Hale, 1979.

Parry K. *Trans Pennine Heritage.* David and Charles, 1980.

Porter J. *The Making of the Central Pennines.* Moorland Publishing, 1980.

Raistrick A. *Old Yorkshire Dales.* David and Charles, 1967.

Raistrick A. *The Pennine Dales.* Eyre Methuen, 1968.

Raistrick A. (ed) *North York Moors.* National Park Guide number 4. HMSO, 1969.

Raistrick A. *West Riding of Yorkshire.* Hodder and Stoughton, 1970.

Raistrick A. *Industrial Archaeology An Historical Survey.* Eyre Methuen, 1972.

Raistrick A. *Green Roads in the mid Pennines.* Moorland Publishing, 1978.

Rollinson W. *A History of Man in the Lake District.* Dent, 1967.

Simmons I.G. (ed) *Yorkshire Dales National Park.* HMSO, 1971.

Slack M. *The Bridges of Lancashire and Yorkshire.* Robert Hale, 1986.

Wainwright A. *Guidebooks to the Lakeland Fells* (7 vols). Michael Joseph.

Wright G.N. *Roads and Trackways of the Yorkshire Dales.* Moorland Publishing, 1985.

MIDLANDS REGION

Cooper B. *Transformation of a Valley, The Derbyshire Derwent.* Heinemann, 1983.

Crump W.B. "Saltways from the Cheshire Wiches." *Transactions of the Lancashire and Cheshire Antiquarian Society,* vol LIV. 1940.

Dodd A.E. and E.M. *Peakland Roads and Trackways.* Moorland Publishing, 1980.

Hey D. *Packmen, Carriers and Packhorse Roads.* Leicester University Press, 1980.

Houghton F.T.S. "Saltways" (from Droitwich). *Transactions of the Birmingham Archaeological Society,* vol LIV. 1929-30.

McKeague P. "Sutton Packhorse Bridge". *Bedfordshire Archaeology,* vol 18.

Mehew S. "Packhorse Bridges in Derbyshire". *Derbyshire Miscellany,* vol I. p.11.

Platts G. *Land and People in Medieval Lincolnshire.* Society for Lincolnshire History and Archaeology, 1985.

Wright N.R. *The Book of Boston.* Barracuda Books, 1986.

SOUTHERN REGION

Bettey J.H. *Wessex from AD 1000.* Longman, 1986.

Crossings Guide to Dartmoor. Reprint of 1912 edn. Peninsula Press, 1990.

Gill C. (ed). *Dartmoor - A New Study.* David and Charles, 1970.

Henderson C. and Coates H. *Old Cornish Bridges and Streams.* Reprint of 1928 edn. D.Bradford Barton Ltd, 1972.

Henderson C. and Jervoise E. *Old Devon Bridges.* A Wheaton and Co, 1938.

Hoskins W.G. (ed). *Dartmoor National Park.* HMSO, 1968.

Kille C.E. "Packhorse Bridges of West Somerset". *Somerset Year Book.* 1933.

Wallis A.J. *Dorset Bridges.* The Abbey Press, 1974.

Printed by CARNMOR PRINT & DESIGN
95-97 LONDON ROAD, PRESTON, LANCASHIRE, UK.

A GUIDE TO THE PACKHORSE BRIDGES OF ENGLAND

Did you know that in Dorsetshire in the reign of George IV, you could be transported for life for damaging a County Bridge? Or that before the Reformation you could earn forty days remission of purgatory if you contributed to bridge maintenance?

For the first time, descriptions of England's packhorse bridges, with accurate locations, are collected in a single guidebook.

A genuine packhorse bridge is defined as being 6ft wide or less, built before 1800 and with known packhorse associations. More than 100 are included. Bridges which are generally described as packhorse bridges but which do not meet these criteria make up a second group of more than 50. A third group of more than 20 bridges have all the attributes of a packhorse bridge, but are without the evidence of packhorse use. Readers of this book are challenged to search these out and decide for themselves.

Introductory chapters describe the history of packhorse transport and the financing and construction of bridges. A County by County guide to 190 bridges follows.

Next time you are out for a walk or holiday in one of our National Parks, make one of these fascinating old bridges your objective.

cp

CICERONE PRESS
MILNTHORPE CUMBRIA ENGLAND

ISBN 1-85284-143-5

9 781852 841430